Fly-Fishing with Leonardo da Vinci

FLY-FISHING

WITH LEONARDO DA VINCI

DAVID LADENSOHN

Terra Firma Books / Trinity University Press

San Antonio

Terra Firma Books, an imprint of Trinity University Press
San Antonio, Texas 78212

Book design by Anne Richmond Boston
Cover images: *The Vitruvian Man*, by Leonardo da Vinci, ca. 1490,
courtesy of Picfair.com; Nicolette AdobeStock
Author photo by Eliza Halley Ladensohn

978-1-59534-305-5 hardcover
978-1-59534-306-2 ebook

Trinity University Press strives to produce its books using methods and
materials in an environmentally sensitive manner. We favor working with
manufacturers that practice sustainable management of all natural resources,
produce paper using recycled stock, and manage forests with the best possible
practices for people, biodiversity, and sustainability. The press is a member
of the Green Press Initiative, a nonprofit program dedicated to supporting
publishers in their efforts to reduce their impacts on endangered forests,
climate change, and forest-dependent communities.

The paper used in this publication meets the minimum requirements of the
American National Standard for Information Sciences—Permanence of Paper
for Printed Library Materials, ANSI 39.48-1992.

Printed in Canada

CIP data on file at the Library of Congress

28 27 26 25 24 | 5 4 3 2 1

For Claudia

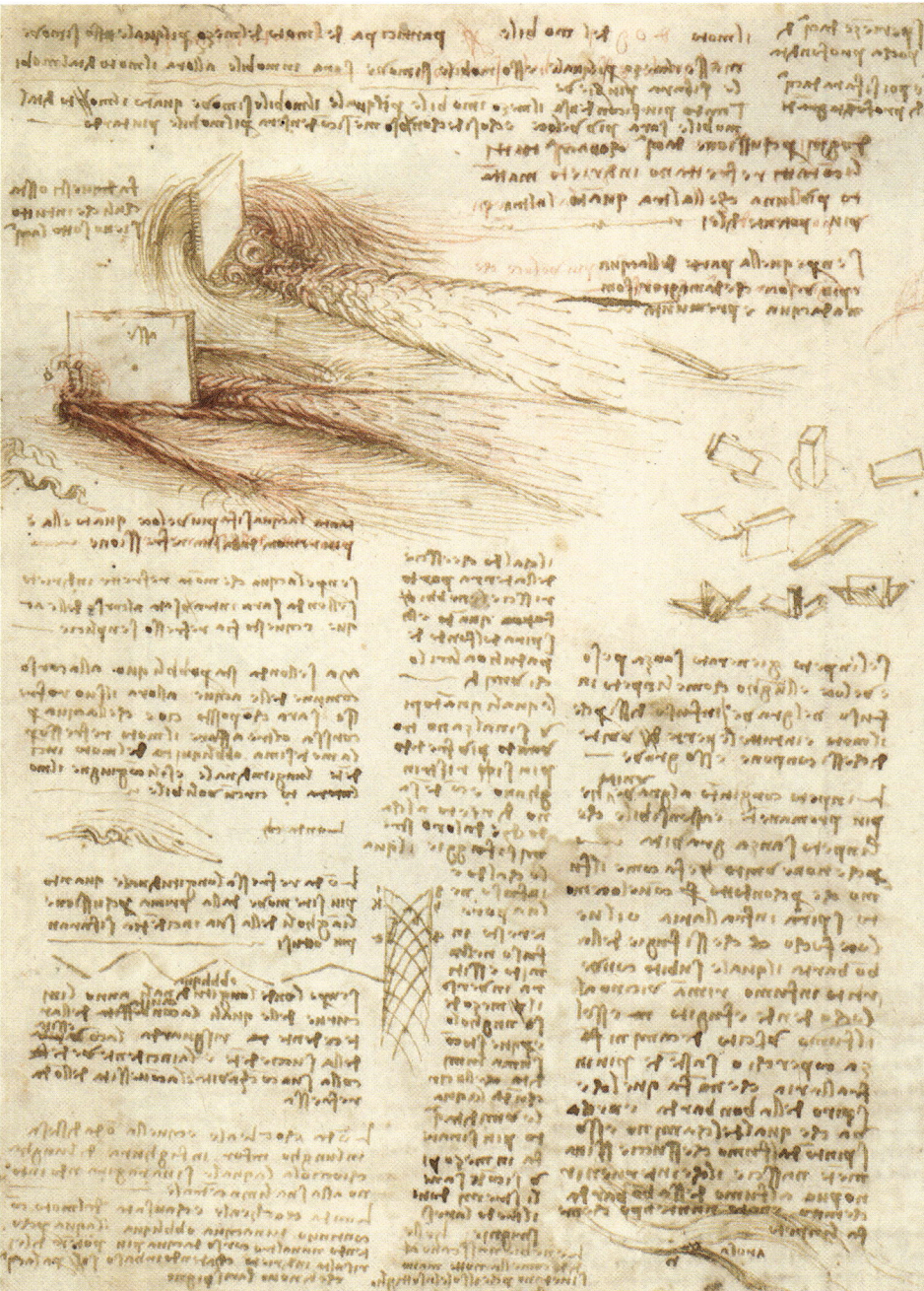

A day well spent brings happy sleep,
so a life well used brings happy death.
— Leonardo da Vinci

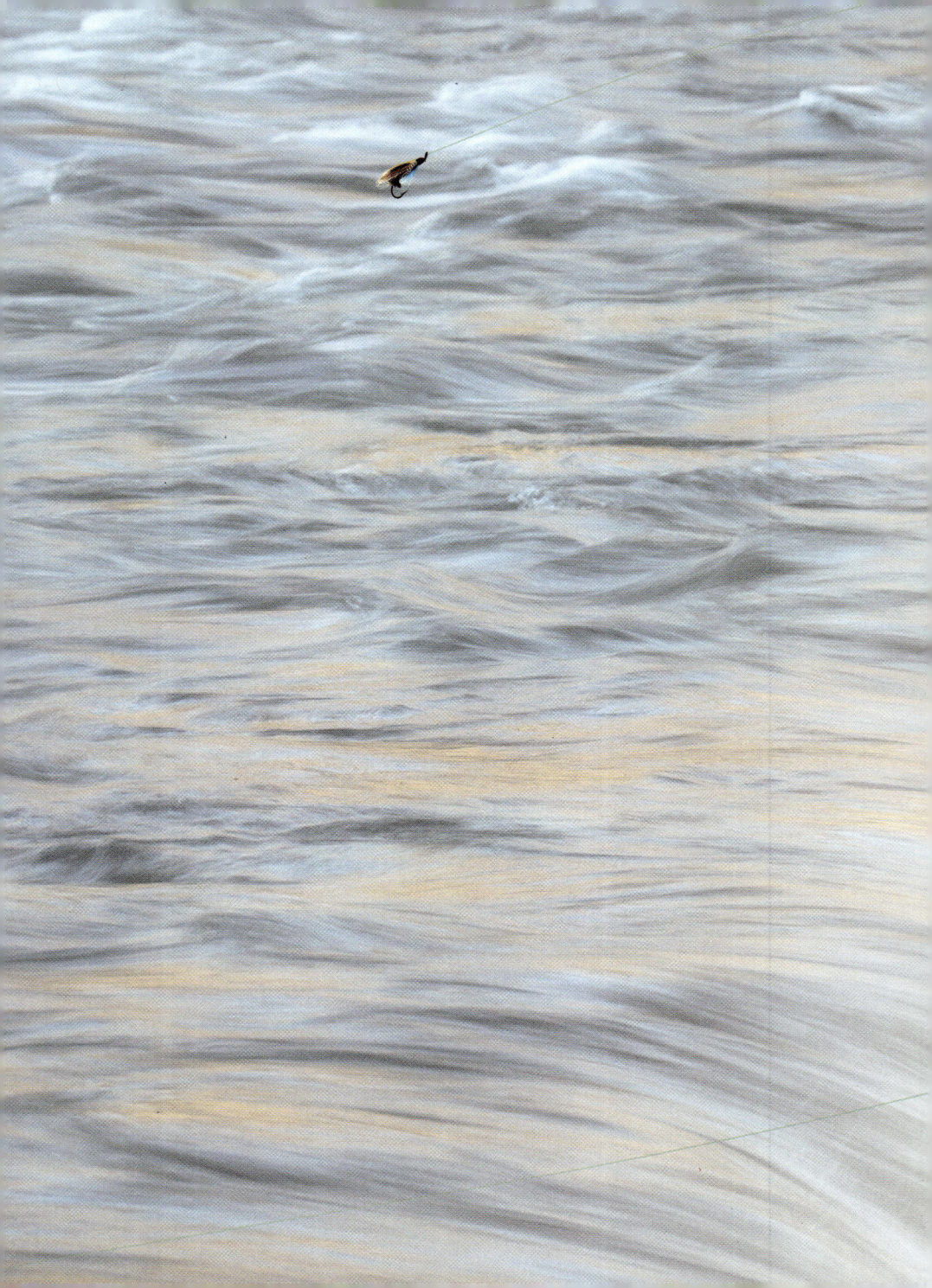

CONTENTS

PREFACE

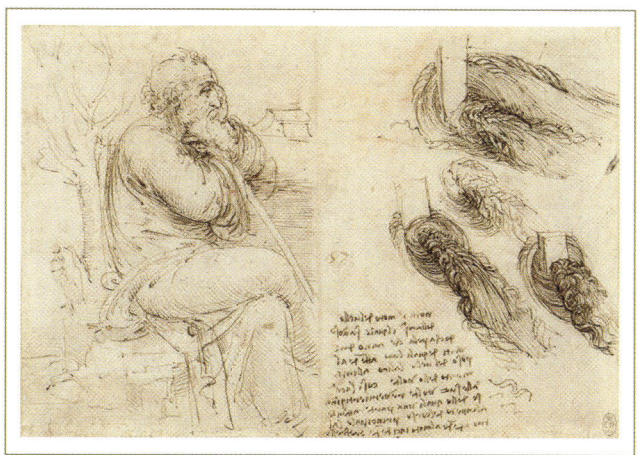

<hr />

<small>"I KNOW EXACTLY WHAT LEONARDO IS THINKING!"</small>

That is what flashed through my mind the moment I first saw this drawing. Then I realized with real embarrassment how outrageous and arrogant that was. Perhaps my brain had been fried by a long day of staring at river currents in Colorado, looking for fish that day; perhaps it was forty years of increasingly obsessive fly-fishing. Leonardo was one of the greatest minds the world has ever known. How could I think I could ever imagine a thought of his? But I did then, and I still do.

In this drawing and many others, Leonardo is trying to figure out how river currents work. Water was an essential element of the larger world he sought to understand, an unstoppable force of creation and destruction. It is the essential element of my small world as a trout angler. Leonardo became obsessed: water is a theme that flows through his entire adult life as an engineer and an artist. His knowledge of rivers—greater than anyone before him— can help anglers solve the enduring puzzle of where to find fish. Learning why he spent decades developing that knowledge can help all of us understand Leonardo the man in a different way.

Since retiring from full-time work fifteen years earlier, I'd had much more time to fly-fish, read, and study water issues in the United States and globally. I did not expect that innocently sitting in our cabin in New Mexico, reading another fine biography by Walter Isaacson, this one about Leonardo, would plunge me into four years of connecting my own disparate worlds through his hundreds of amazing water drawings. I did not expect to find Leonardo, the lover of all natural things, and Leonardo, the hydraulic engineer, behind Leonardo, the famous painter. I certainly did not expect to find myself inside Windsor Castle studying his own studies, or to bring Leonardo's perspective back with me to every stream I visit. But I catch more fish now, and I have a richer life in the process.

INTRODUCTION

I WAS ONLY A FEW HOURS OFF THE PLANE from Boston to London, but the adrenaline of meeting this legend gave me the energy to ignore how far out of my depth I already was. Much of what I knew about Leonardo had come from this man's many books, written over five decades.

There was no one else in the small restaurant north of Oxford when he walked in, slender and wearing his signature high-collared vest. Fortunately, Martin Kemp, the best known of the world's Leonardo scholars, is a gracious man, just as he had been in correspondence during the previous months. Not even five minutes after we sat down for dinner, our conversation was skipping from topic to topic, century to century, from art to science and back again, at Kemp's lightning speed. He talked about the downsides of specialization, which took modern intellectuals so far from the fertile ground Leonardo had been able to plow in every direction.

We were in a quiet booth in the back of the place, and the crowd I had expected never materialized. No doubt the wine and jet lag relaxed me, so I finally popped the question I had carefully prepared: "What would it be like to have dinner with Leonardo da Vinci?"

Kemp's eyes lit up, and he launched into his answer like he was swinging at a fat pitch. Clearly my cleverness was not novel; he had thought about this and likely answered it many times before. The insights he shared that evening were fascinating and of a piece with what other generous scholars would tell me when I subsequently met with them. Leonardo experts have lived inside the master's head for years because they have read his notebooks, written over a lifetime for himself alone. Each is part book draft, part sketchbook, part money account book, part reflection, part to-do list, and only a small part diary, regrettably.

The experts described the Leonardo they know as well as they know their close friends, a man who had shared with them his frustrations, his joys, his unbounded curiosity and intellectual vigor. Those discussions eventually led me to ask myself, What would it be like to go fly-fishing with Leonardo da Vinci as my guide?

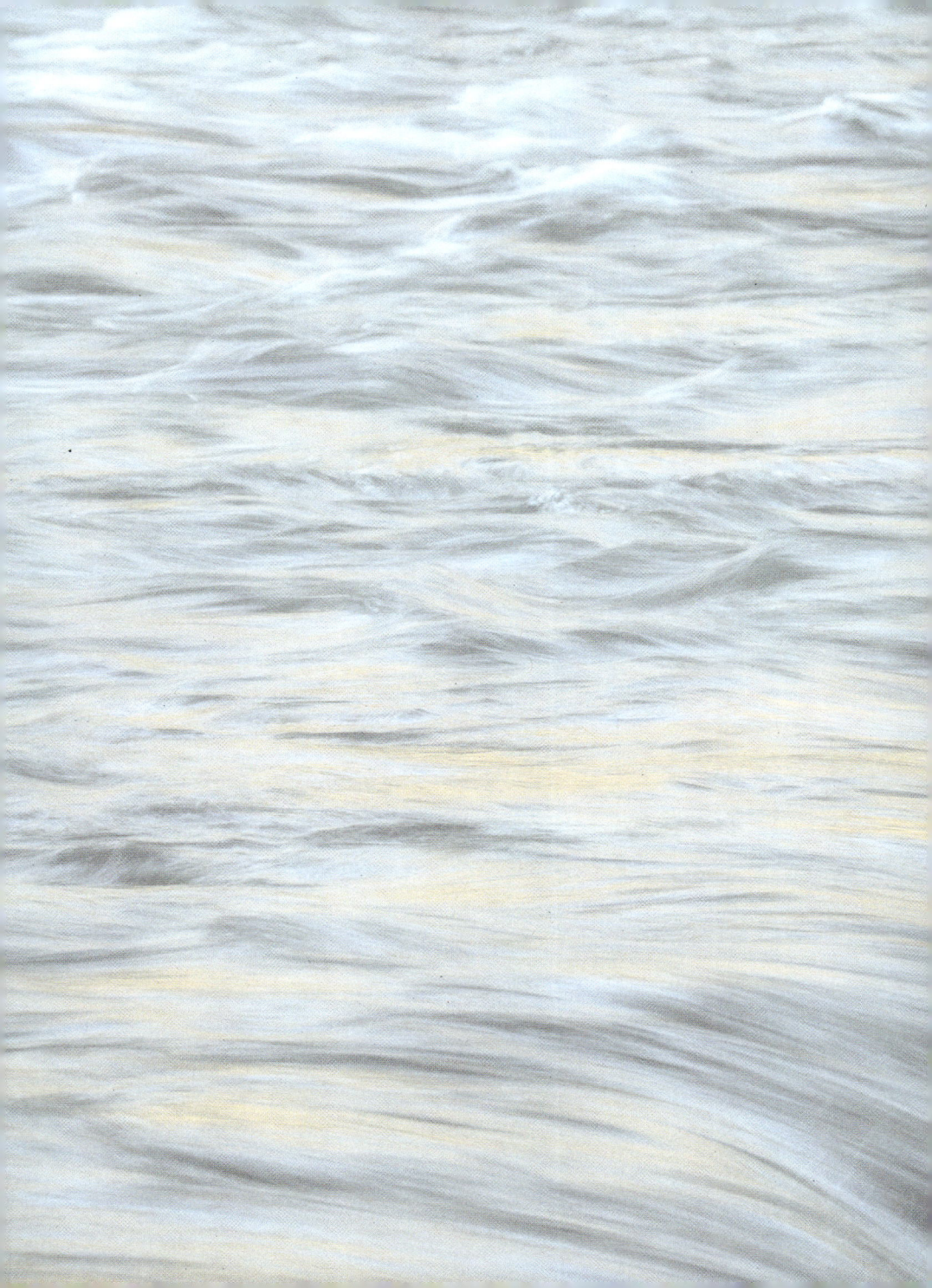

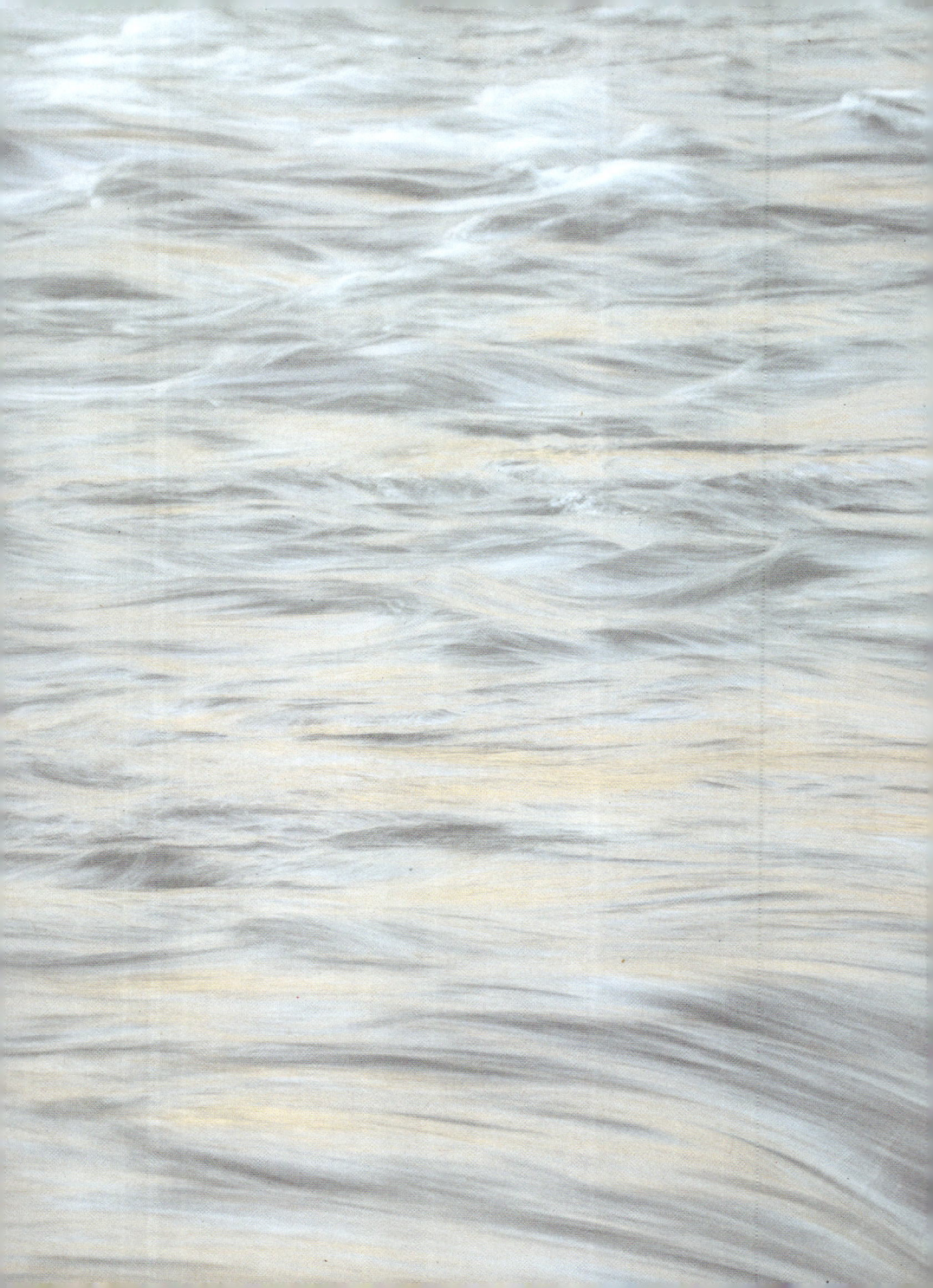

WHO'S THE LUCKY BASTARD NOW?

IT DID NOT BEGIN WELL. In July 1451 an up-and-coming twenty-four-year-old lawyer had a roll in the hay with an orphaned peasant girl of fifteen. It happened in a small town in the hills of Tuscany, eighteen miles west of Florence, called Vinci. His name was Piero, and hers Caterina. Whether it was actually just a fling or an ongoing young love affair is unknown, but we do know that they had a child out of wedlock. The little boy was likely born in a small stone cottage not far from where Piero's father lived comfortably. That is not to say, however, that Ser Piero da Vinci's love child with Caterina di Meo Lippi was a secret.

The day after he was born, on April 15, 1452, the infant was baptized in the church in Vinci as Leonardo di Ser Piero da Vinci. Piero's own god-father served as little Leonardo's godfather, along with nine other godparents. Leonardo's eighty-year-old grandfather proudly recorded the birth of this first grandchild, the first son of his firstborn son, and his social peers attended the happy event.

If there was any awkwardness in Leonardo's illegitimacy, it would have come from the fact that by the time Leonardo was born, Piero was already engaged to another teenager, a sixteen-year-old woman of his own social class. They would be married by the end of the year. After Piero stayed close by his newborn child for a week, he went back to work at his office in the center of Florence, near the seat of government. Soon after he found Caterina a suitable husband, a local farmer and kiln worker with whom she would have five more children. Caterina and her husband remained on good terms with Ser Piero for at least the next twenty years. (If she was actually a Jewish slave girl, as recently proposed by one scholar, Piero was the one who drew up the surviving document that freed her from bondage.)

Piero and his new wife, Albiera, never had children of their own. Piero continued to attend to Leonardo somewhat. By the time the boy was five, Piero moved him from Caterina's home, where she was tending to her younger children, to Piero's father's large home on Via Roma near the center of Vinci. Grandfather Antonio seemed to enjoy relaxing and collecting rents on his farmlands more than actually working. He and his youngest son, Leonardo's bachelor uncle Francesco, were kind and loving to the child. These three generations of leisure males got along well. Leonardo had the benefits of one household with a mother and numerous children, a second home with his indulgent grandparents and uncle, and occasionally a third home in the big city where his father and kindly stepmother now lived.

What Leonardo did not have was legitimacy.

This is where our luck and Leonardo's luck come into the picture. Had he been born within a proper marriage, or even legally adopted by Ser Piero, he would have been the sixth-generation first son to be automatically entitled to inherit property and, more importantly, automatically entitled and expected to become a notary. He would have been admitted to the guild of notaries in Florence, which functioned a bit like both a religious club and an exclusionary union, and he would have received the honorific title "Ser" before his name, like his notary father and great-grandfathers. But that was not to be. The guild did not accept *non legittimo* (bastards).

So, Leonardo was destined instead to become a mere craftsman, clerk, or artisan to earn a living. When he was twelve, both his beloved grandfather in Vinci and his stepmother in Florence died. Ser Piero brought Leonardo to live with him in the city. At this point, with his hopes of a legitimate heir from Albiera dashed, Ser Piero could have legitimized Leonardo and attempted to get around the rules banning him from the guild. He did not. Instead, he enrolled Leonardo for a few months of second-rate "abacus schooling," where future tradesmen and bank clerks learned rudimentary mathematics and bookkeeping. Teaching was in Tuscan dialect. Leonardo missed out on the multiple years of the first-rate "Latin schooling" a future notarial wordsmith would have received. Then Ser Piero delivered the young teenager to a painting workshop to become an apprentice, something beneath the dignity of a firstborn legitimate son. But for all the rest of us lucky bastards, it was a decision that would reverberate with importance for the next five hundred years.

TROUT ARE COUCH POTATOES

ONE OF THE FIRST THINGS I HEARD from a guide when I started fly-fishing forty years ago is that trout want to eat the maximum amount of food while expending the minimum amount of energy. Well, who doesn't? This, we are told, is why they hang out in "seams," where two currents with different speeds meet. The faster water brings a conveyor belt of bugs right past a spot where the trout can lazily hold in the slower current. The fish will move briefly into the faster water, gobble a passing morsel, and quickly move back onto the couch nearby. Other places that satisfy this condition are in slow water behind a rock or the cushion of water in front of a rock, where current bounces back against current, creating a calm place for a fish to "sit" or "hold," waiting for food to come by. A log in the water creates a slow place behind it and faster water off the end; a shallow riffle of rocks usually ends in a deeper, slower trough where fish can lie, waiting for the bugs, which lose their grip on the stones and wash into the current downstream. This also works vertically: trout can hold happily in a slow current below a fast current, which is much more difficult for the angler to perceive, unless they are with a guide who has a great understanding of hydraulics.

After food, the second of the Big Three needs of trout is oxygen. Trout do not breathe air but rather pull dissolved oxygen out of the water as the water flows over their blood-red gills. Two effective ways to unintentionally kill a trout by depriving it of oxygen are to hold it out of the water longer than you can hold your breath underwater, or to fight it past exhaustion in water that is warmer than sixty-seven degrees. Conversely, cold water splashing over rocks or riffles or waterfalls, large and small, holds lots of oxygen to keep trout spunky.

(For those who are unfamiliar with fly-fishing terminology, please see the fly-fishing jargon primer at the end of the book, which explains some of the most commonly used terms.)

The third important need for trout is to avoid getting eaten—they need protection. The biggest mortal danger comes from birds, particularly ospreys, herons, mergansers, and kingfishers. Trout need a place to hide from these predators, like under a log, a low-hanging shade tree, or an undercut riverbank. Expert Tom Rosenbauer points out that fish don't always hold right in the hiding place, but they need to be close enough to one to get there quickly when they sense a threat. Most often, that threat is telegraphed by a moving shadow, which might be an osprey overhead or might be a fly line overhead; the trout does not analyze the source before sprinting for cover.

Were Leonardo a fishing guide, he would have easily held the solution to the most important part of the trout puzzle: where to find the fish. A poor cast with the wrong fly will certainly do better in a honey hole full of fish than a good cast in places with no fish. But we can do better than that. Guides talk a lot about "presentation," putting the artificial fly in front of the fish in a way that looks most like a real bug. The "real" is less about the exact shape, size, and color of the fly than about its behavior in the water. A natural insect will drift along with the current, on or in the water. Leonardo cast tiny grass seeds onto currents to see exactly what the waters did with them, and we can do that too. A real bug will not produce little wakes, nor skate diagonally across the water, nor whip by the trout's face twice as fast as other bugs and debris. But those kinds of "drag" do happen when the artificial fly is tied to a fly line that is lying across multiple currents between the angler and the fish's "feeding lane."

I like to ask guides what they think the most important pieces of gear and most important skills are for all fly-fishers. The gear answers I get frequently are a good pair of polarized sunglasses to see the fish and the fly and a good floating line. For skills, guides usually answer (1) an angler's ability to manage their line and (2) their ability to read the water. Guides never say that it is most important to have an expensive rod or reel, or to be able to cast fifty or sixty feet—the very things anglers talk about most.

"Line management" is a lot of geometry. The best analogy is the children's game of crack the whip, where the kid at the end of the line ends up having to move much faster than those in the front who are running around in random directions. As the faster river current between the rod tip and the fly pushes the middle of the fly line downstream into a curved "belly," the fly at the end drags rapidly across the slow current like a child at the end of the whip, in a way no freely floating natural insect ever would. Managing the line encompasses all the things we do to correct for these unnatural effects of the currents on the fly, in the right ways, in the right amounts, at the right moments. Although people said Leonardo was not good at math, I believe this was meant in comparison to mathematicians of his day, not to the rest of us. His notebooks are filled with complex geometric work aimed at understanding the effects of light and shadow on objects both earthly and lunar, and at understanding human vision. His fine motor skills and coordination, his knowledge of human musculature and geometry, combined with his musician's sense of timing, would have made demonstrating and teaching line management to fly-fishers natural for him.

≈

Humans have an infinite imagination to make up challenges. We challenge each other in games and make bets about who can throw the farthest, run or skate the fastest, jump the highest. We seem to do it for status, for fun, for entertaining ourselves and others, and for profit by both the contestants and those who bet on them. (Betting on skill and outcomes is not modern; the ancient ball game played by the Aztec, Maya, and other Mesoamerican cultures three thousand years ago saw people betting money, the clothes off their backs, and even their own servitude! Happily, the stories about the losing team's players getting sacrificed may well be legend, and we definitely don't sacrifice anglers after they lose a fish.) These challenges and games go far beyond physical sports into crossword puzzles, memory challenges, chess, go, speedreading, poker, and on and on.

Humans like to make the challenges ever harder. Playing the card game of hearts leads to spades, which leads to bridge. Checkers players take up chess, and soccer rules preclude using our very useful hands. Rifle hunters become bow hunters. While in life and business, we seek the easiest and most cost-efficient way to accomplish a task, with our leisure time, we seem to do the opposite.

And so it is with fishing. If you want to catch a fish, the most efficient and successful way is to bait a multipronged, barbed hook with a worm, a live shrimp, a piece of corn or cheese, or a marshmallow. When anglers choose to substitute a rubber worm or plastic lure, they make the entire sport more difficult, craftier, cleverer, and much more expensive for themselves. Then we reduce the strength of the rod and the thickness of the line in order to put more difficulty into the equation, to require more finesse and skill. We think of it as leveling the playing field with the fish.

Fishing with flies instead of bait or lures is like playing golf from the championship tees. The casting is more difficult because the weight is not at the end of the line, as it is with all other throwing activities. The flies themselves are either too wind resistant to cast easily or too tiny to see on the water. We mash down the barb of the hook to make it easier for the fish to slip off if we don't keep a tight line. We use lighter, flimsier rods and tippets (the thinnest portion of the line), which are easier for the fish to break, partly to make them less visible to the fish but also to challenge ourselves to use technique more than strength in playing and netting a fish, if we're lucky enough to hook one.

And we are very proud of ourselves! Sometimes a bit too proud, looking down our noses at other types of fishing and anglers, smug in imagining ourselves the rightful descendants of British angling gentry.

The modern fly-fisher accumulates endless amounts of the latest equipment: rods, reels, special tools, flies, and the many accoutrements that adorn their vests and packs. Like amateur tennis players and golfers, we believe that such gear will greatly improve our skill, when practice and logic are actually more important. Leonardo would argue that observation and thinking are the most important things to carry to the stream. "Knowledge is the captain," he wrote, "and practice, the soldiers."

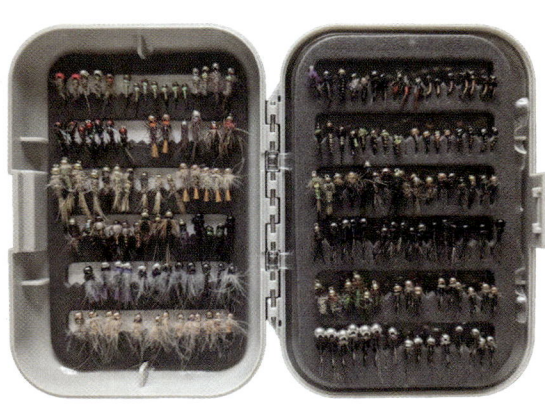

"I need a different fly"

I'm as guilty as anyone of carrying unnecessary numbers of fly patterns to the river. We joke that flies are really tied to attract the fisher, not the fish. I probably carry fifty to a hundred flies at any given time, whereas one-tenth that number will serve a good angler well. "One fly" tournaments challenge participants to pick one specimen of one pattern and fish that single fly all day, neither choosing poorly nor losing it in a fish or a tree.

Figuring out where the trout is and how to get a fly to her pays the largest dividends. Figuring out where to stand probably comes second. Both are far less expensive than rods, reels, and boots.

Leonardo made fun of materialism in his notebooks, even as he also sought financial security and fame. Here he skewers those who spend their energies accumulating wealth and possessions and titles instead of reveling in the natural gifts given to all people: "It is ordained that to the ambitious, who derive no satisfaction from the gifts of life and the beauty of the world, life shall be a cause of suffering, and they shall possess neither the profit nor the beauty of the world."

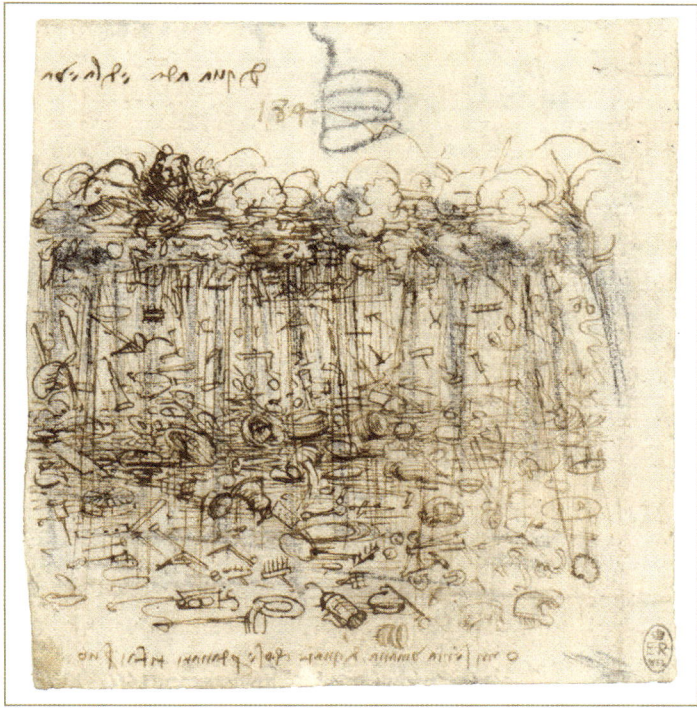

Leonardo, *A Cloudburst of Material Possessions*, 1506–12

His drawing *A Cloudburst of Material Possessions* is both telling and full of humor. At the bottom he wrote: "Oh human misery, how many things you must serve for money." Was this directed at everyone else, or was Leonardo complaining about his own situation? He knew the alternative and strove for it:

> [Whatever can be lost should not be called riches;] Virtue is our true good and the true reward of its possessor. That cannot be lost; that never deserts us, [until] life leaves us. As to property and external riches, hold them with trembling; they often leave their possessor in contempt and mocked . . . for having lost them.

Leonardo managed his resources carefully, but he did not dress in sackcloth nor take vows of poverty. His condemnation here is of those who ambitiously chase only wealth of the material kind, who do not stop to smell the roses—or perhaps fish in beautiful mountain streams.

Countless friends have asked me why I love to fish. Sometimes I tell them that trout live in beautiful places; sometimes I describe the many facets of the puzzle that each bend in the river presents. But in the end, I think that the mystery of each cast is the heart of it: will this one produce a strike, a tug and fight, a wild fish to hold for a moment and then release?

Fishing guides tell their clients to "think like a fish." While their intentions are good, this advice is ridiculous. How can I think like a fish? Maybe like a chimp or gorilla, to whom I am distantly related. Maybe even like a whale or dolphin, species that seem to think, reason, communicate, and play in ways somewhat similar to humans. But a fish? With a brain the size of a pea? It's insulting. And embarrassing that they so often outsmart us!

When we talk about what a fish thinks, we are actually reasoning backward. Experienced guides and anglers have noticed some sort of pattern in what fish do under some circumstance. Then they imagine why a human would do such a thing and impute that motive to trout. We say, "They think your shadow is a bird, a predator, so they hide." "They think that a 'streamer' is a wounded minnow floating downstream and easy for them to catch and eat." "They think a Prince Nymph" (which resembles absolutely nothing) "looks halfway like a pale mayfly 'emerger' and halfway like a stonefly larva." In science and logic, this is labeled inductive reasoning: working backward from the specific example to the general case. Deciphering fish thoughts is likely working backward at least one bridge too far.

When a guide or experienced angler tells a novice what fish think, the novice will tell her buddy what fish think. The buddy will tell his friend. By the time we eventually hear what fish think, it is received wisdom handed down from experts. Since it comes from people who are more experienced than ourselves, we are likely to believe this conjecture to be fact.

Leonardo would cry "Bologna!" on all of this. He would notice when fish behave as predicted and when they do not. And he would likely say that keeping your shadow from falling onto trout usually works better for sneaking up on them than the opposite, and leave what they "think" out of it. He would observe, experience, and make deductions. If his experience did not match what the scholars or fishing books or guides said, he would trust his own experience and observations more.

Guide John Belozer says it best: "If fish were that smart, they'd be fishing for us!"

THE APPRENTICE

WE DON'T REALLY KNOW MANY DETAILS of Leonardo's childhood from the time he moved in with his grandparents and uncle at age five until the time the newly widowed Ser Piero brought Leonardo to live with him in Florence at age twelve. Living those formative years in Vinci, first with and then near his half-siblings, probably gave him a "gang" to run with in the fields outside the village proper. I fantasize that they might have fished with worms or cheese in the little streams outside Vinci and Anchiano, had swordfights with sticks, and just been kids. Since he was the only child of the house, it seems likely that he was indulged by both his grandparents and his uncle Franco, just fifteen years older than Leonardo. His grandmother's family, in addition to producing its own long line of notaries, owned a pottery factory. Leonardo's drawing may have started there; it almost certainly would have been encouraged by his family and the factory's artists.

As noted earlier, Ser Piero enrolled Leonardo in an abacus school in Florence to learn basic business math, reading, and writing. This formal education probably lasted less than a year, perhaps only a few months. Those headed for a profession went to a "Latin school" for years. Leonardo, the boy, would carry a feeling of educational inferiority into adulthood, sometimes referring to himself as an "unlettered man" because of his lack of ability to read books in Latin. When he was about thirteen, his father apprenticed him to Andrea del Verrocchio, one of the leading painters and sculptors in Florence. Verrocchio was a client of Ser Piero and was supposedly impressed with some of Leonardo's drawings, which Ser Piero showed to him. An apprentice in those years was fed and housed at his master's workshop, did as instructed, learned some specific skills in the process, and eventually took up the craft he had learned. He might eventually join the guild of masters in that industry after the apprenticeship. Apprentices fairly often took the family name of their master, but Leonardo kept his own father's name. Sometimes the family of the apprentice paid the master to take on the child, in addition to the free labor provided.

Florence in the 1460s, when Leonardo was a teenage apprentice, was a very different place than we experience as tourists centuries later. This thriving city had a population of only about fifty thousand, approximately the size of Bozeman, Montana, or Ithaca, New York, today. Florence was still recovering from the plague, or "black death," which had arrived in 1348. In just four years, plague had killed half the people of Florence and surrounding Tuscany. The entire Republic of Florence, with about 750,000 people in Leonardo's day, was much smaller than the powerful Republic of Venice. The Papal States, including Rome, were larger still, as were the independent Kingdom of Naples and the Duchy of Milan. But throughout the preceding century of devastation, Florence had remained the cultural center of the Italian peninsula.

Young Leonardo's Florence was humming with artists and commerce and money. Its namesake *florin* was the most trusted gold coinage in Europe. Nonetheless, there were haves and have-nots. Although his father was a professional from a family with land—the only real form of capital for centuries—Leonardo was among the have-nots. With no wealth, no serious education, no legitimate right to inherit, and talent only for a mere craft (artists did not have professional cachet during most of his lifetime), Leonardo was headed for a career making products that the haves might or might not buy.

Verrocchio's workshop was one of the finest in Florence, but still one of dozens. Employees and apprentices in such places would make objects from metal or wood, or paint religious scenes to order or sell from inventory. The master would instruct them, paint the most difficult or critical parts of a scene, and seek commissions from wealthy patrons for art to display in their homes or contribute to their churches. Workshops were part studio, part factory, and part retail store. The most important commission for Verrocchio in Leonardo's early years was the enormous gold-covered ball, topped by a cross, to crown Brunelleschi's dome of Florence's cathedral.

Imagine Leonardo in this place. Not the Leonardo of fifty years later, quietly concentrating in his private rooms, adding a small brushstroke to the still unfinished *Mona Lisa*, but the teenage Leonardo in a bustling workplace. Boys making paintbrushes and grinding mineral powders into pigments for older, full-fledged assistants. Noisy sawing and sanding of wooden panels being prepared for painting, and the swearing if any sawdust or dirt got onto a master's freshly painted surface. Cutting, hammering, and soldering of the copper gores that would form the great ball. Careful guarding of valuable gold and silver foils that would be applied to the surface of base metals. And the buzz of instructions, orders, coordination, shoptalk, and gossip, punctuated with the occasional laugh or angry outburst of a senior worker berating an apprentice.

We do not know when or why, in these early years, Leonardo became interested in water and river currents. Was it the attraction of peace and quiet, which contrasted with the workshop? Or a reminder of the countryside he may have missed? Could it have been related to some childhood trauma? When he was only four, a rare and destructive hurricane rose in the Adriatic

Sea east of Italy, crossed the peninsula, and struck the Arno Valley outside of Florence, near Vinci. The winds and rain must have been terrifying and might have left a lasting impression on the mind of even a child so young.

The hurricane is described emotionally in an otherwise long and sober *History of Florence* written after Leonardo's death by his younger friend Niccolò Machiavelli. A brilliant official of the Medici period, Machiavelli gets a bad rap today for his dispassionate and clear-sighted book on governance, *The Prince*. After his fall from influence and power in Florence, imprisonment, and torture, Machiavelli was exiled to his nearby farm but was later officially commissioned to write the city's history. Because the storm occurred before Machiavelli was born and is so vividly described, as if by an eyewitness, biographer Serge Bramly speculates that his source may have been Leonardo himself.

When Leonardo was thirteen, a large flood struck Florence. When the Arno burst its banks in 1465, bridges and people would have been swept away, not to mention livestock and possessions. Witnessing the city's inundation, loss of life, and damage to property must have been shocking. The hurricane of Leonardo's childhood and the major flood of his teenage years made clear to him the destructive power of water, well beyond the power of humans to resist. Did they contribute to the apocalyptic *Deluge* drawings he produced more than fifty years later? Or to his desire to design better structures to limit the damage? He wrote that "among . . . destructive terrors, the inundations caused by impetuous rivers ought to be set before every other awful and terrifying source of injury. But in what tongue or with what words am I to express or describe the awful ruin, the inconceivable and pitiless havoc, wrought by the deluges of ravening rivers, against which no human resource can avail?"

If Leonardo understood water's destructive violence, he also knew the beauty and serenity of the streams around Vinci and the central importance of the normally peaceful Arno:

*Now [water] brings a conflagration, then it extinguishes; is warm and is cold;
now it carries away, then it sets down, now it hollows out, then it raises up,
now it tears down, then it establishes, now it fills up and then it empties . . .
now it speeds and then lies still, now it is the cause of life and then of death,
now of production and then of privation . . . and now with great floods it
submerges the wide valleys.*

And: "So at times it is turbulent and goes raving in fury, at times clear and tranquil it flows playfully with gentle course among fresh meadows."

Leonardo incorporated these two opposing faces of water into his thinking, his art, his scientific studies, and his entire career. He would seek to understand water in its beauty and mystery, its life-giving qualities, its shaping of the earth, and its power to destroy mountains, farms, cities, and their people. He would use this original knowledge in countless ways.

His earliest surviving drawing is of the Arno flowing gently through a valley in the countryside, sketched when he was twenty-one.

Perhaps Leonardo's early and ongoing obsession with water came simply from two truths he observed: "Without it nothing can exist among us," and "water is the source of all life." In his future, lifelong quest to understand everything about everything and to formulate a grand unifying theory of nature, water may have been the very best thread to flow through it.

Leonardo did well as a teenage apprentice in Florence, learned quickly, and was soon given opportunities to display his skill by painting some of the lesser figures and landscapes around the central figures, which Verrocchio or another senior member of the workshop would finish. Leonardo's abilities eventually proved so good that legend claims Verrocchio swore off painting ever again, once his student surpassed him. At age twenty, Leonardo was admitted to the painters' guild as a master. Lest we think of a master in too lofty terms, it really meant only that he had completed his apprenticeship and

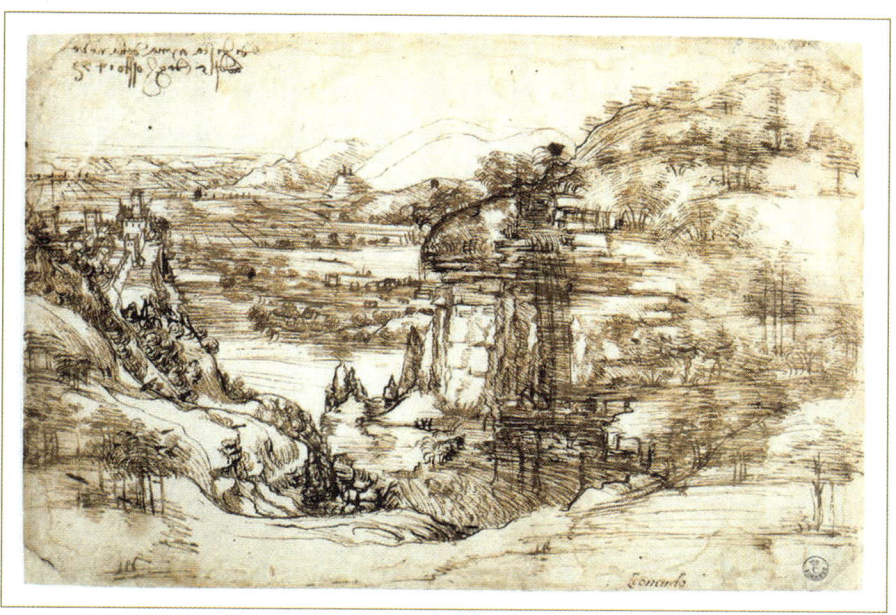

Leonardo, *The Arno Valley on the Day of St. Mary of the Snow*, 1473.
Leonardo's oldest surviving drawing was an innovation, a pure landscape
with no human figures to tell a story, no historical event to celebrate,
and no moral instruction implied.

could take on apprentices and open his own workshop if he wished. He apparently did not wish; he continued to work out of Verrocchio's workshop, now as an assistant being paid for his efforts. And the guild was only a part of the Company of Physicians, Apothecaries, and Grocers, more of an association for celebrating religious occasions and speaking with one voice than a select academy or society.

About this time, Leonardo assisted Verrocchio on the painting *The Baptism of Christ*. Art historians point to the painting of the angel in the lower

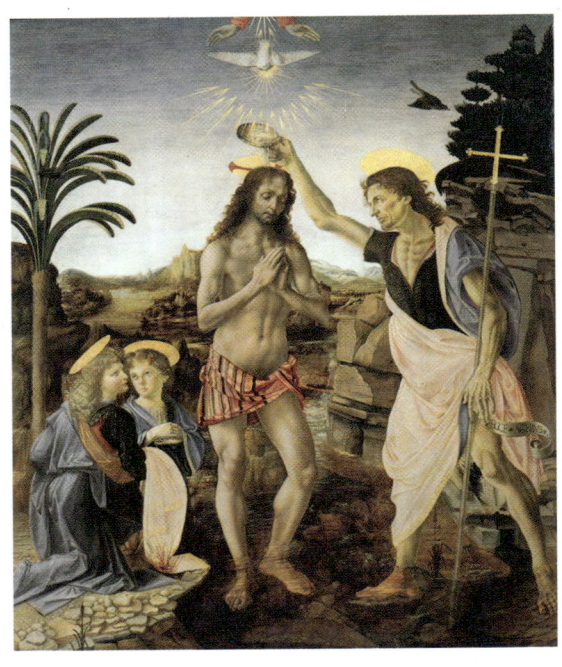

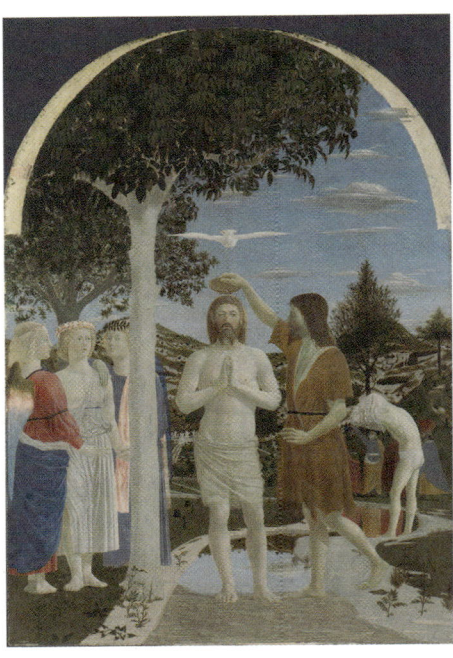

Verrocchio/Leonardo,
The Baptism of Christ, 1472/75,
detail below

Piero della Francesca,
The Baptism of Christ, 1450s,
detail below

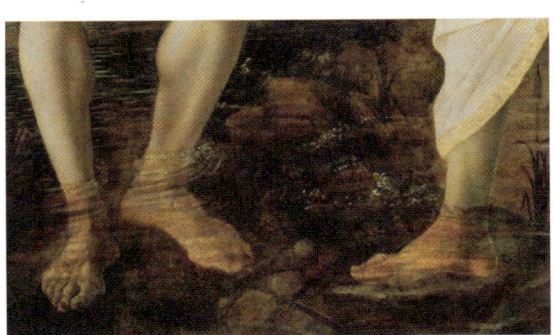

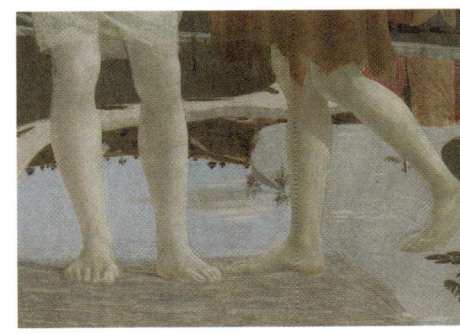

left as superior to the other figures and, therefore, clearly Leonardo's work. What interests me, though, is the extraordinary rendering of the water swirling around the ankles of Jesus and John the Baptist. The same scene portrayed by contemporaries in both Italy and Germany shows none of the understanding Leonardo already exhibits of how river currents actually flow past obstacles. At most, other artists merely portrayed that water in a river makes things appear darker; but Leonardo understood what the River Jordan's current would do when it bumped into, or "percussed," a solid leg. It would be another ten years before his scientific and engineering studies of water began in earnest, but he had clearly observed the currents of the Arno and small streams near Vinci and figured out how to render his knowledge in paint.

One realistic detail that Leonardo omitted in *The Baptism of Christ* is the refraction of objects in water. Martin Kemp told me that it would be "indecorous," or bad taste, to portray Jesus and Saint John in this accurate but distracting manner. So, too, the cross John holds.

Since Leonardo was the keenest of observers and also becoming a serious student of optics, there is no doubt that he was well aware of refraction and its illusions. In sight-fishing, it is important to know before casting that a fish you see under the water is farther away than it appears; its image is refracted and the angler must adjust for it.

After reading an early book on optics that had been translated into Italian, Leonardo wrote: "Experience does not err, but rather your judgements err when they hope to exact effects that are not within their power."

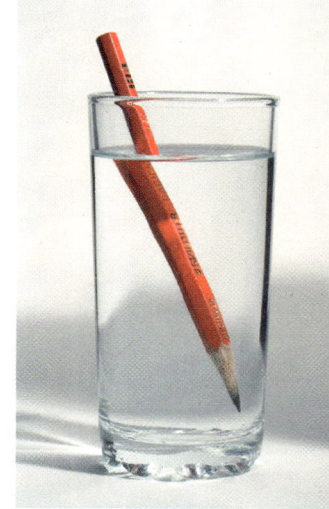

Refraction

This statement is not merely a comment about illusions but a broader judgment Leonardo had reached regarding science. To understand the laws of nature means to understand what is possible and impossible. Alchemy and perpetual motion were avidly pursued in Leonardo's time. His beliefs that it was "not within their power" for alchemists to find ways to turn base metals into gold, or for engineers to create perpetual motion machines, were likely minority views. Leonardo would seek certainty through observation and experimentation, not through received wisdom or popular opinion.

After completing his apprenticeship with Verrocchio, Leonardo also worked in the studios of other prominent masters in Florence like Uccello and the Pollaiuolo brothers. He studied the work of the still-living Leon Battista Alberti, an engineer, mathematician, architect, and philosopher, like the future Leonardo. These were all intellectual men who studied anatomy and geometry and would have stimulated Leonardo's curiosity in all areas. Things were going well for Leonardo; his talent was proving stronger than his illegitimacy and lack of education.

Then he was arrested.

WHAT NOW?

JUST ONE WEEK BEFORE HIS TWENTY-FOURTH BIRTHDAY, Leonardo and three other young men were accused of sodomy with a male prostitute. The penalties for such a crime were serious, including fines, imprisonment, expulsion from Florence, or even death. He left us no diary or letter to help us know what impact this had on his own view of his career, future, and life. He must have been mortified by embarrassment, terrified of prison. We can only imagine and empathize.

Leonardo's sexual orientation is not really in doubt, although it was not much discussed before the mid-twentieth century. Today more scholars debate whether the *Mona Lisa* depicts Mona (My Lady) Lisa Gioconda than dispute whether Leonardo was gay. And while few would have disputed Leonardo's homosexuality during his lifetime, the prevailing attitude in Renaissance Florence seems to have resembled "Don't ask, don't tell."

Homosexuality was fairly common in Florence. Many of the leading artists, including Verrocchio, Donatello, and even the city's late hero-architect, Filippo Brunelleschi, were likely gay. German slang at the time for gay men was "Florenzers." While this might imply that homosexuality was no problem in Renaissance Italy, that is far from the truth.

Boxes were placed around the city specifically for people to report crimes of gay acts to Florence's special Office of the Night. Leonardo's name found its way into one of these boxes. The accusation was a public matter; Leonardo and the three other men may have even been locked up briefly. But then Leonardo got very lucky. The anonymous accuser never came forward as was required, so the hearing on the charge was first delayed for two months and ultimately dropped. Probably this was not just luck: one of Leonardo's codefendants was related by marriage to the ruling Medici family. There are many ways in which the accuser may have been "encouraged" by the powerful Medici to not testify.

Nonetheless, the arrest and accusation were a lasting black mark for the rising young artist and cannot have helped either his confidence or his relationship with his father, whose prominence as a notary was growing.

Leonardo spent six more years in Florence. He opened his own workshop but received only two commissions, both from clients of his father. He failed to complete either project, even though he had agreed to clear contractual obligations. In Verrocchio's workshop, Medici patronage had supported everyone. Now Leonardo was stagnating while contemporaries like Sandro Botticelli and Filippo Lippi were thriving. Biographers Giorgio Vasari and Walter Isaacson each suggest that Leonardo suffered from depression sometime during this period. Although he was studying mechanics, such as the equipment Brunelleschi designed to lift his giant dome atop the Duomo, the scientific light was not yet shining brightly. It was time for a major change of scenery.

THE WIT AND WISDOM OF A FISHING GUIDE

NICK STREIT IS A MOUNTAIN MAN. He is big, with a big voice, big red hair, and a big red beard. He brings protein for his family from the mountains, shooting elk and deer with a bow and arrow, no gun. He butchers animals up there and carries them for miles on his back to the truck. His wife, Chrissy, hunts, too, as well as raises rabbits, chickens, and turkeys in their backyard, which looks out on yet more mountains, south of Taos, New Mexico. They hunt wild mushrooms in the forest, and Chrissy cooks all of this into hearty meals for their two engaging children; all four of them are avid readers.

Nick's dad, Taylor, was my first fishing guide in the Southern Rockies, back when I was still in my forties. A handsome ex-hippie refugee from New York, Taylor is sardonic, telling me he needed to move out of Taos because it had "too many crystals." Taos is one of the great countercultural towns in America as well as a remarkable mixing place of Indian, Spanish, Anglo, cowboy, and nouveau Western second-home people and cultures. It has a large colony of "earthships" in the desert west of town—energy- and water-efficient houses built mainly underground out of tires (filled with compacted dirt), concrete, and glass. Their intensely passive solar design maintains a temperature of seventy degrees through snowy winters and blistering summers.

I had called Taylor in 1996 because I found his name in a book about fly-fishing in northern New Mexico. He'd written the chapter on fishing around Taos, and my wife was trying to get me to go along with her idea of buying a small cabin in the mountains, using fly-fishing as the bait. I was totally against the project but promised to be open-minded. Over the phone, I asked this stranger where to look for a cabin around Taos, and he answered in a way I later found to be typical. "If I wanted to look around Taos, I wouldn't look around Taos. I'd look around Chama." Almost thirty years later, we are still happily living in a cabin in the woods outside Chama, population 1,300.

After many years of Taylor's expert advice (well worth taking: he was inducted into the national Fresh Water Fishing Hall of Fame as a "Legendary Guide"), he told me one day that I should fish with his son, because "Nick is really a better guide." I cynically assumed he said this because Nick had reopened Taylor's old fly shop and probably needed the business. But when I started fishing with Nick, I found some truth in what Taylor had said. The big difference: Taylor likes to be alone and Nick likes people. A good guide must like people enough to put up with their botched casts, their missed fish, their inability to understand or execute what the guide is telling them, and their long, repetitive stories (guilty!).

When Nick was a schoolchild, Taylor moved them to the Bahamas, where he could explore the fishing and guiding opportunities there. They lived as squatters in someone's unfinished house on the beach. Nick attended local schools and absorbed the difficult lessons of being in the tiny racial minority. Later, father and son spent the harsh New Mexico winters exploring the fish-filled rivers of summertime Patagonia, driving to large estancias (ranches) and simply asking for permission to fish their owners' private waters. Although quiet, Taylor is a raconteur and the author of excellent books on fishing, and he tells some of the stories from those days in one titled *Man vs. Fish*.

Nick does not get lost in the mountains. Like Leonardo, he has exceptional powers of observation and makes landmarks out of trees and rocks that, to me, quickly look like all the other trees and rocks. He used to guide me in tame places a mile or so from the pavement. Now I see a date with Nick on my calendar as a promise to explore remote country I would never dare try on my own. Even then, he has led me to a few great fishing spots that we had to swim or climb vertically from to leave because, it turns out, he hadn't actually been there before.

The worst six words a client can hear from a guide are "You should have been here yesterday." It means that you are having a bad day, whereas the angler previously on the same beat cleaned up. Or it may mean that the river is great and your abilities are the problem—and the guide is being charitable.

Fly-fishers spend many hours with their guides; good guides become good friends. Good fishing guides need patience, wisdom, and wit. Patience to make you, the client, not feel bad when you miss a fish because you set the hook too late for the fourth time in a row. Patience when you get the guide's last perfect fly hung up in the same tree—again—just out of her reach. The wisdom to calm you down and add perspective to both bad days and good days. And the wit to keep you entertained. Leonardo had each and every one of these great-guide qualities. A conversation could be factual or philosophical: "The water you touch in a river is the last of that which has passed and the first of that which is coming. Thus it is with the present." And he understood rivers. Isaacson says, "Leonardo recorded 730 findings about the flow of water" in his notebook that we call the "Codex Leicester" and elsewhere "listed sixty-seven words that describe different types of moving water."

Leonardo had the patience to study the complexity of river currents his entire adult life and never gave up trying to work out the laws governing their sometimes turbulent and chaotic patterns. (Little did he know that it couldn't be done completely, even with supercomputers.) He patiently worked on the *Mona Lisa* for sixteen years without ever putting his final brushstrokes on the panel. He sued his half-brothers over his only family inheritance and then agreed to leave the contested property to them in his will. Leonardo's patience was often confused by others with a sheer inability to finish projects, but he certainly never gave up on an idea or a friend. And a good fishing guide does not give up on a client's inability to show improvement if the client demonstrates a sincere desire to improve and a willingness to listen.

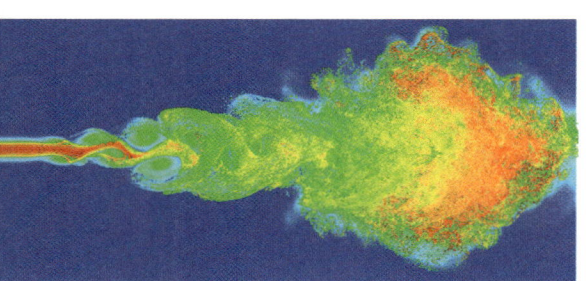

Computer modeling of the transition from laminar, or smooth, currents to turbulent, or chaotic, currents. Leonardo hoped to understand the process and use accurate predictions for his engineering.
It remains an unsolved problem.

Good fishing guides need wit; Leonardo's wit was legendary. He once wanted to study and portray the details of laughter on men's faces, the muscles and crinkles that anatomically reveal what physically happens when they laugh. So, he sat down with a group of drinkers and told them jokes until they were all laughing hysterically and next captured the myriad physical details in his mind's eye, his spectacular visual memory. Afterward he went back to his drawing table and recorded all of it in his sketchbooks.

He made up scores of jokes, fables, and allegories about people, animals, plants, and even rocks to express his thoughts about life and human foibles to entertain his listeners. He wrote them down, presumably to practice them so they could be told with the special timing comedians and storytellers appear to exhibit effortlessly. It would have been a deep pleasure to spend a day on the river with Leonardo, laughing or trying to make up your own jokes and riddles to keep the banter going. He would have made a slow day of fishing much less disappointing, as great guides have done for me. He might recite, as he wrote: "It was asked of a painter why, since he made such beautiful figures . . . , his children were so ugly; to which the painter replied that he made his pictures by day and his children by night."

Leonardo made up many humorous riddles for his listeners to solve during festivals at court in Milan. He stated them in the form of prophecies, often ominous or crazy-sounding, such as these:

That shall be revered and honored and its precepts shall be listened to with reverence and love, which was at first despised and mangled and tortured with many different blows. (Answer: Paper for wise writing is originally made from beaten rags.)

The bones of the dead shall be seen by their rapid movement to govern the fortunes of their mover. (Answer: Dice used in games of chance.)

Many persons puffing out a breath with too much haste will thereby lose their sight and, soon after, all consciousness. (Answer: Blowing out the candle before going to bed.)

Good guides also need wisdom. Practical wisdom for fishing and for staying safe in remote places. And general wisdom for discussing with clients who get reflective while out in nature, or as they see life's seasons passing by. Leonardo wrote down hundreds of serious observations about the nature of different sorts of people: lazy ones and active ones; intellectually curious ones like himself and avaricious ones whose time spent chasing after ever more wealth or power he believed was a foolish waste of their lives. When an angler develops trust in her fishing guide and vice versa, discussions can get serious about religion and politics, those taboo subjects I was taught in taxi driver school to never broach with a client. If Leonardo ever trusted someone's confidence enough, he might have revealed philosophical thoughts, too, although there is no specific evidence that he ever revealed much of his deepest thoughts to anyone other than his own notebooks. But here are a few of those: "He who wishes to be rich in a day will be hanged in a year." And, "You can have no dominion greater or less than that over yourself." Also, "Patience serves us against insults precisely as clothes do against the cold. For if you multiply your garments as the cold increases, that cold cannot hurt you; in the same way increase your patience under great offences, and they cannot hurt your feelings."

Those fishing guides who tell stories to clients over a long day on the river also need the social intelligence to know what is enough and what is too much:

Words which do not satisfy the ear of the hearer weary him or vex him, and the symptoms of this you will often see in such hearers in their frequent yawns; you therefore, who speak before men whose good will you desire, when you see such an excess of fatigue, abridge your speech, or change your discourse; and if you do otherwise, then instead of the favor you desire, you will get dislike and hostility.

And if you would see in what a man takes pleasure, without hearing him speak, change the subject of your discourse in talking to him, and when you presently see him intent, without yawning or wrinkling his brow, . . . you may be certain that the matter of which you are speaking is such as is agreeable to him.

Leonardo's ability to entertain was also reflected in his music and his production of pageants and performances by others for the amusement of princes' courts and visitors. These were unique because he combined traditional spoken or sung entertainment with marvelous stage sets, including sound effects and devices of his own invention, like mechanical birds and actors suspended, flying through the air by way of machinery he also invented, and elaborate costumes of his design.

I mentioned Leonardo's musical ability, perhaps an unusual talent to find paired with a serious engineer and scientist. He was described in his own time as having a beautiful singing voice and the talent to play the lyre well. (No guide has yet broken into song for me on the river. Thom Chacon, the one professional singer who ever guided me, said he never would either.) This lyre was not the instrument we associate with angels or the young King David's psalms. Rather, it was a type of fiddle called a *lire de braccio* in Italian, with some strings to be played with a bow and some to be plucked. Leonardo won a contest for his playing in Florence and accompanied his own singing of classic poems and improvised, humorous songs.

But all this was to come later.

ORIGINALS

SESSLER'S BOOKSTORE ON WALNUT STREET in downtown Philadelphia already felt ancient back in 1971. It was a long and narrow space where little of the natural light that came in through the wood-framed windows in front reached the back. It sold new books, but what fascinated Claudia and me were the old books, neatly organized, with leather bindings and titles stamped in gold on the spines. We were just nineteen years old, college students, hardly imagining that we would eventually be married for fifty years, living around the world, and getting old together. What was old back then was Miss Mabel Zahn, way in the back of Sessler's, at her large desk. She was small and wizened and wore her gray hair pulled back in a bun. We did not realize it at the time, but she had started work at Sessler's in 1905 at age fifteen and had been president of the shop since 1955.

We visited that store often but, being on student budgets, seldom bought anything. Miss Zahn must have perceived a genuine interest, though, because she let us look at eighteenth- and nineteenth-century science books with hand-colored illustrations of butterflies and rare early twentieth-century books on modern art. One day she initiated the conversation and was clearly excited. "Let me show you something I've just acquired—a letter written by George Washington from Valley Forge." She handed us the five-page letter to look at and handle. She pointed out passages relating to the miserable conditions in that winter of 1777–78, conditions that helped shape the disciplined character of the Continental Army, which would fight on for four more years before winning the new nation's independence. (Severe lack of food, blankets, and shoes was the main problem; the winter was not as bitterly cold as later myth portrays it.)

I had never handled a rare, important, original document. That such a thing even existed outside a museum had probably never occurred to me, and it was fascinating—there was an instant feeling of a direct link to history. That feeling has never left me.

During the 1970s when I was a banker, I began regularly going to Washington, DC, on business. I always added a few hours to visit the Library of Congress in its old building. In the basement, the librarians would call up the earliest printed atlases for me to pore over, original maps printed in 1482 and 1508 and 1513, based on the second-century *Geography* of Ptolemy. Coincidentally, these were all by-products of the Byzantine scholars bringing ancient Greek and Arabic manuscripts and knowledge to Italy, and the spread of printing during Leonardo's lifetime—sparks of the Renaissance. During the 1990s, when I ran a company manufacturing barracks furniture, I frequently visited navy bases in Chicago and discovered the wonderful Newberry Library there. It, too, had every edition of the Ptolemy atlases, and it, too, allowed me to study the originals in its high-ceilinged, quiet reading rooms.

Miss Zahn, the Library of Congress, the Newberry. These generous people and institutions contributed to my waking up one day in 2021, after studying Leonardo's water drawings in books for two and a half years, and knowing that it was time to see the originals. ◄—

MASTER OF WATER

LEONARDO NEEDED A JOB. He was almost thirty years old, with his own apprentices to grind pigments and prepare papers for drawing. He had ideas and more experience; he also seems to have regained his ambition and energy after his possible depression. But he had no steady income, no patron, meager savings, and a less than excellent reputation in Florence. He had probably lost his best client due to his failure to complete a commission, and the politics in Florence had become extremely messy. Worst of all, the Medici ruler had just sent six outstanding Florentine artists to help decorate the new Sistine Chapel at the pope's request; Leonardo was passed over. Leonardo da Vinci needed greener pastures.

The duke of Milan, to the north, seemed a good prospect for employment in 1482. Milan was truly the Big City, with more than one hundred thousand people—double or triple the size of Florence. Leonardo had an artist friend there who would take him into both his home and his workshop.

Leonardo had visited Milan that year on behalf of Florence's ruling Medici, Lorenzo the Magnificent, as part of a diplomatic mission to reduce tensions between the two city-states. (The fifteenth-century city-states of the

Italian peninsula were almost constantly warring with one another, making and breaking alliances and inviting the French and others into their armed disputes.) Leonardo had performed for the court, singing and playing his arm-lyre (*lira di bracchia*) and a lute he had sculpted from silver in the shape of a horse or cow's skull. Lorenzo reportedly sent the lute as a gift for the duke, who was likely impressed with both the gift and the singer.

Both the duchy of Milan and its "duke," Ludovico Sforza, could use a bit more of Florence's cutting-edge culture to buff up their reputations. People called Ludovico "Il Moro"—the Moor—as if he were from North Africa or Spain—because of his dark complexion in this fair-skinned part of Italy. Ludovico's father had grabbed power through force, and Ludovico, the same age as Leonardo, could be pretty ruthless himself. He was a man of contradictions. On one hand, he was highly educated and interested in all of the arts, which he both practiced himself and promoted at the Sforza court. On the other, he fought for five years over the power to control his nephew, the legitimate duke, now twelve years old. He had just banished the boy's mother and taken power as regent a year before Leonardo came to Milan. Ludovico eventually may have poisoned his nephew to get to the top. A major program of expensive cultural spending like Florence's might put some lipstick on the pig that was Il Moro's ducal court.

But Milan needed more than just culture when Leonardo arrived. It was at war with Venice and other city-states on the Italian peninsula; Il Moro needed military talent. Notwithstanding a principled dislike of war and the lust for power, Leonardo eventually sent him a job application, written from left to right by a scribe in highly legible script. The entire letter is a marvel of puffery, even by modern résumé standards and even for someone as genuinely skilled in so many fields as Leonardo. In hindsight, the letter is also humorous since only in the last of his eleven bold claims to expertise in many other fields does Leonardo da Vinci mention that he can "also" paint quite well:

Most illustrious Lord, Having now sufficiently considered the specimens of all those who proclaim themselves skilled contrivers of instruments of war, and that the invention and operation of the said instruments are nothing different from those in common use: I shall endeavor, without prejudice to anyone else, to explain myself to your Excellency, showing your Lordship my secrets, and then offering them to your best pleasure and approbation to work with effect at opportune moments as well as all those things which, in part, shall be briefly noted below.

1. *I have a sort of extremely light and strong bridges, adapted to be most easily carried, and with them you may pursue, and at any time flee from the enemy; and others, secure and indestructible by fire and battle, easy and convenient to lift and place. Also methods of burning and destroying those of the enemy.*

2. *I know how, when a place is besieged, to take the water out of the trenches, and make endless variety of bridges, and covered ways and ladders, and other machines pertaining to such expeditions. [. . .]*

5. *I have means by secret and tortuous mines (tunnels) and ways, made without noise, to reach a designated [spot], even if it were needed to pass under a trench or a river.*

6. *I will make covered chariots, safe and unattackable, which, entering among the enemy with their artillery, there is no body of men so great but they would break them. And behind these, infantry could follow quite unhurt and without any hindrance. [. . .]*

10. *In time of peace I believe I can give perfect satisfaction and to the equal of any other in architecture and the composition of buildings public and private; and in guiding water from one place to another.*

Item: I can carry out sculpture in marble, bronze, or clay, and also in painting whatever may be done, as well as any other, be he whom he may.

Again, the bronze horse may be taken in hand, which is to be to the
immortal glory and eternal honor of the prince your father of happy memory,
and of the illustrious house of Sforza.

And if any of the above-named things seem to anyone to be impossible
or not feasible, I am most ready to make the experiment in your park, or in
whatever place may please your Excellency—to whom I commend myself with
the utmost humility, etc.

The duke clearly needed military engineers as much as painters, and Leonardo opportunistically offered to be one, with serious attention to water in addition to fortifications and weaponry. He claimed to have unique designs for militarily useful, strong, light bridges, to know how to remove water from trenches, possibly including moats, and to know secret ways to tunnel under rivers. He also claimed to have peacetime skills "in guiding water from one place to another," an important claim since Milan expanded its economic and cultural reach by the extensive building of canals for transportation and irrigation. All of these professed skills require deep knowledge of how water destroys man's structures. Leonardo had already focused on river currents.

Leonardo got hired, sort of. Although he did not immediately receive a monthly stipend and some financial security, he began to get work. In addition to a commission to paint a portrait of Il Moro's beautiful mistress, he was hired to produce theatrical entertainments for the ducal court. Other commissions followed, including for the fresco of *The Last Supper.*

Importantly, Leonardo got opportunities to work with water experts.

It is hard for us to imagine today what rivers meant to life during the Renaissance. Everything we do on divided highways and in the skies, from personal travel to commerce to international trade and air freight, was conducted primarily on boats plying rivers, canals, and seas. These were the

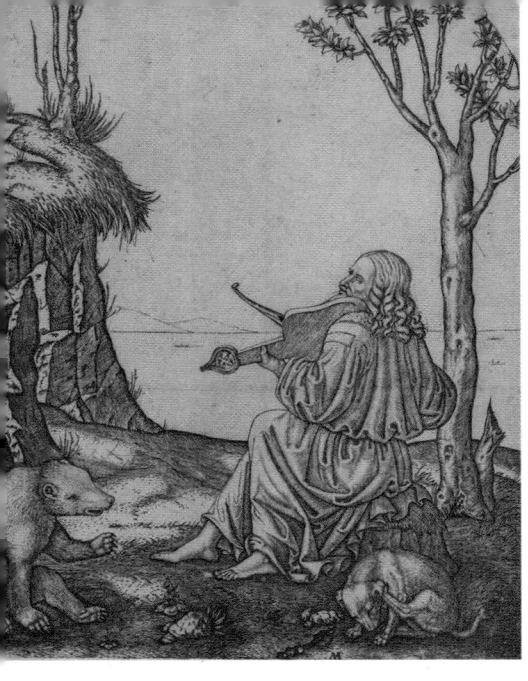

Marcantonio Raimondi, *Orpheus Charming the Animals*, ca. 1508. Orpheus was normally depicted as a clean-shaven youth. Music historian Ross Duffin believes this engraving may portray a middle-aged Leonardo playing his *lira da braccio*.

lifelines of cities, duchies, and republics. Water for humans, animals, crops, and goods was both a vital resource and an enormous challenge; flooding, erosion, and navigability all had to be managed. The winding, unruly, flood-prone Arno through Florence and the network of canals in and out of Milan were essential to the city-states' existence.

Rivers and canals were also the electric grid of Leonardo's day. They powered mills that ground flour for bread and pasta, mills for sawing timber and marble, and mills that spun wool and silk for weaving cloth, the most important export industry of bustling Milan. Boats used the canals to carry fiber and fabric out to the world and to bring marble from the mountain quarries to build the magnificent cathedral that is Milan's soaring landmark to this day. Maintaining, extending, and improving those waterways and controlling their floodwaters was vital in peace and war, vital to the movement of goods, people, and ideas. They made Milan an inland port, and they made Milan wealthy.

It would take time and study to become recognized as a *maestro d'acqua* (master of water), a professional consulting hydraulic engineer for a government. Efficiency and expansion of the canal system centered on Milan were important, so hydraulic engineers were important. Leonardo made much of his living at this occupation and at military engineering, probably as much or more than he did from artistic commissions (when he actually finished those). This water work also drew perfectly on his curiosity and scientific, engineering, architectural, aesthetic, and artistic sensibilities. As icing on the cake, being employed so variously for engineering, theatrical, and artistic pursuits offered him time to pursue his scientific investigations and an occasional place at court, where his desires for lasting fame and financial security might eventually be satisfied.

The Renaissance man that Leonardo represents to us today emerged during his seventeen years in Milan. His skills as a painter, an expert in river currents, in engineering, in creating stage sets and spectacles for entertainment, his fascination with flight and light and shadow and so much more, developed as he matured there. He turned talent into expertise through tireless observation, deep thought, and bold experimentation. His curiosity was insatiable, and his ability to connect everything to everything else was unprecedented. And one of those things was how he could read the water.

Classic pocket water, where boulders or logs break up the stream
and create pockets of slower currents and more depth for trout to hold

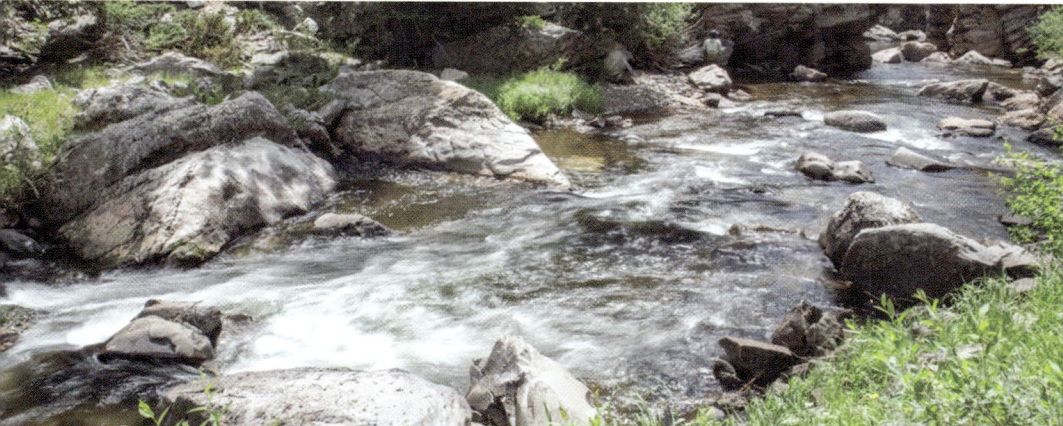

READING THE WATER

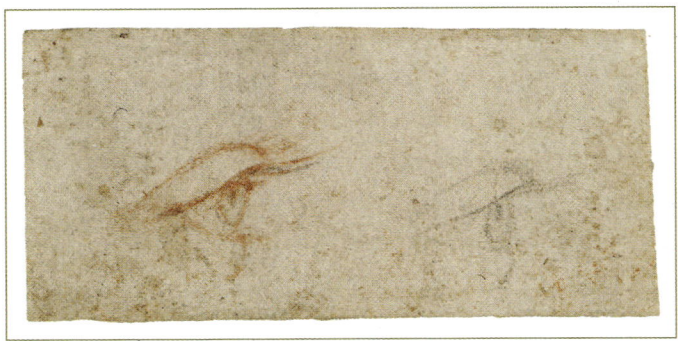

The mastery of its course is sometimes on the surface, sometimes in the center, sometimes on the bottom. One portion rises over the transverse course of another, and but for this the surfaces of the running waters would be without undulations. Every small obstacle whether on its bank or in its bed will be the cause of the falling away of the bank or bed opposite to it.

— Leonardo

ALTHOUGH WE MAKE FLY-FISHING FOR TROUT REMARKABLY COMPLEX, one simple principle stands above all others: you can't catch fish where there are no fish. The most elegant, long cast with the perfect fly will not produce a strike if there are no fish there. Famous bank robber Willie Sutton is apocryphally credited with saying that he robbed banks because "that's where the money is." Anglers say 90 percent of the river doesn't hold fish; we need to spend all of our time fishing the 10 percent. "Reading the water" is figuring out which is that 10 percent. It starts with looking for places with food near holding water, oxygen, and cover.

Leonardo lived before chemistry became a science, explaining how water captures and dissolves oxygen. He may or may not have observed how fish stayed near cover and places like "bubble lines," which steadily bring insects down the river. But these two requirements are logical and easily explained. The most complex element of finding fish is to understand currents; Leonardo understood currents better than anyone.

I imagine describing to Leonardo what makes the ideal places for fish, what I call "Goldilocks"—places with that perfect combination of fast current for food, slow current for fish to hold in, plenty of oxygen, and a hiding place when necessary. He would probably respond with some contemporary Tuscan dialect version of "Got it" and proceed to walk upstream, pointing out spot after spot that met the conditions. His superb powers of observation would be multiplied by his unique understanding of currents, particularly subsurface currents, to identify the very best places to fish.

If you learn what Leonardo knew, you will catch a lot more fish. How did he learn? Through careful, constant observation: "All our knowledge has its origin in our perception." Five hundred years later, an American philosopher offered a similar lesson: "You can observe a lot by just watching" (Yogi Berra).

Most anglers cast indiscriminately, searching a river for fish. Guide Stefano Sabbatelli, a member of the Italian national fishing team and former coach of the American team that became world champions, told me he has made as few as two or three casts in an entire day! Very careful observation is required to follow Nick Streit's overarching maxim, "Fish with intention."

Leonardo studied the works of previous hydraulic engineers on canals from Lombardy around Milan, Tuscany around Florence, and around Venice, far to the east. He learned what was effective in using river currents to resist floods and erosion. He studied locks that allowed navigation upstream and downstream where the natural fall and rapids of rivers were too great for oarsmen to overcome. He proposed new canals and designed the miter gate for canal locks, a brilliant innovation that is still used on the Panama Canal. He thoroughly applied his deep intelligence and intense observation to understand how currents work both at the surface and deeper in the "water column." Thinking about those hidden currents is essential for any angler using nymphs or streamers to catch the majority of fish that feed subsurface, where they are safely out of view of birds, their chief predators.

"Does this river always flow the same direction?" Nick tells me he gets this bizarre question about once every year. Of course, rivers always flow downhill, toward the sea. Except when they don't—and that is an important exception, because an eddy is frequently an ideal place to find a trout, often a large one.

When the main current of a river flows around an obstacle like a boulder or a log with some speed, the water pressure is greater on the front and sides than it is behind that rock. You will feel this immediately if you stand in the current in front of or beside a boulder. The area behind the boulder or log now has lower water pressure, and the "outside" water will flow in there to fill the void. In the way that air flowing from high-pressure cells into low-pressure cells causes winds, this backward flow of river water creates a current independent of the main river current. It is an eddy.

Some eddies are tiny, like the little whirlpools behind a canoe paddle. Some are bigger, like those behind a boulder or a projecting point of the riverbank. And some are huge, where half a river turns around to flow upstream for a hundred yards or more. Some eddies back up slowly, and some spin so fast they can suck a trout fly, or a whole boat, under the water. Leonardo was obsessed with eddies and spinning vortices. Fly-fishers should be too.

Leonardo discovered the "conservation of flow"—the concept that the same amount of water passes every point in a river, fast or slow, shallow or deep. The water of an eddy slows down or even becomes almost still. The low pressure and the vortex pull food in from the main current and put it on a slow-moving serving platter for the trout to inspect and eat, all while demanding little effort from them to resist the faster main current. Eddies can be such a choice spot to feed that large fish claim the territory and chase out smaller interlopers. Eddies can also accumulate bubbles and dirt and leaves, trapping insects in a foamy surface scum that fish can cruise through and hide under from birds and anglers. Twice I have spent more than thirty minutes with my face less than three feet from a large trout while it swam slowly against the gentle current of the eddy, hoovering up midges and other insects circulating with the current. I believe the trout were so focused on their feast that they could not be distracted by me. Sneaking up on eddies on your knees, slowly and quietly, can be a great fishing tactic.

Casting into a foam-covered eddy can be fun when a fish visibly moves foam in the taking of a dry fly. But it can be tricky when any movement of an "indicator" or dry fly in the foam is hard to see, and the trout eating the nymph below gets away undetected.

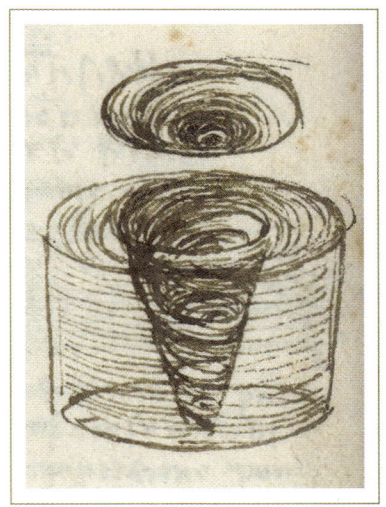

One of Leonardo's many drawings of vortices, which fascinated him his whole life (detail)

I do not know when or why Leonardo first became fascinated with eddies and vortices. I have wondered if he saw a large and frightening one that day of the hurricane when he was four, or whether someone scared him as a child with a story about getting sucked into a giant whirlpool. In the *Deluge* drawings made late in life, he depicted wind, water, and lightning destroying people, houses, horses, trees, and entire towns. He wrote about the need for artists, including himself, to draw and paint these frightening scenes.

A more likely explanation than some childhood trauma is that eddies are unusual and Leonardo wanted to understand them. Whirlpools have a hypnotic attraction, like the flames of a fire, and they have a scientific attraction: what creates them, and how do they work?

There was also a fascinating scientific anomaly for him to study: "The spiral or rotary movement of every liquid is swifter in proportion as it is nearer to the center of its revolution. This is a fact worthy of note, since movement in a wheel is so much slower as it is nearer to the center of the revolving object."

Many of Leonardo's most interesting drawings were his way of observing currents percussing obstacles and reasoning his way through what was happening and why—drawing out loud: "Observe the motion of the surface of the water which resembles that of hair and has two motions, of which one goes on with the flow of the surface, the other forms the lines of the eddies; thus, the water forms eddying whirlpools one part of which are due to the impetus of the principal current and the other to the incidental motion and return flow."

Leonardo's 1504 study of two weirs on the Arno for the government of Florence (pages 46–47) traces the currents and vortices that so fascinated him. Note the many small vortices he observed next to the large ones. These are important when we think about where trout hold in rivers. To the right, downstream, the vortices begin to dissipate and a less turbulent main current takes form again. Trout would be most likely to hold in this calmer area and in the seams where the main current meets the waters flowing in slowly from the eddies.

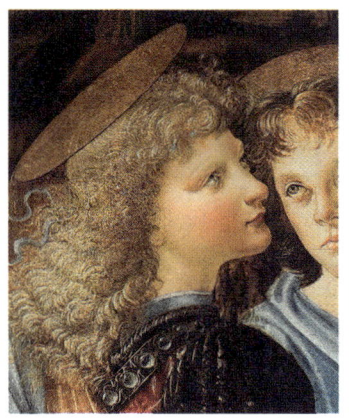

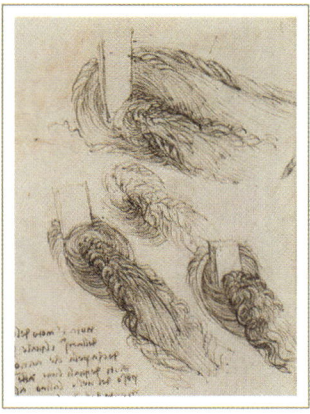

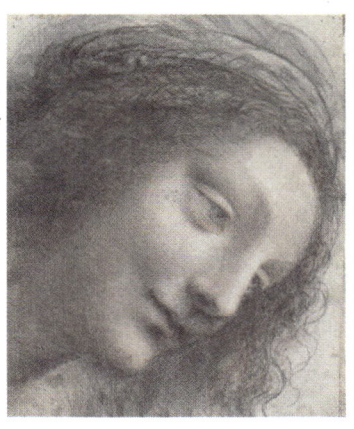

The angel in this detail from Verrocchio's *Baptism of Christ* already shows Leonardo painting hair with the same enthusiasm and style he has for river currents.

Detail of the drawing (page xi) adjacent to Leonardo's note comparing water in motion to hair

Riverlike hair details, Leonardo, *The Head of the Virgin*, 1510–13

At the bottom center of *A Weir on the Arno East of Florence*, you can see where the waters had eroded the river's wall and bank, first for at least 110 feet and then again for more than a hundred yards. Such erosion was common, threatening walls, houses, mills, and farmers' land holdings. Leonardo wanted to apply his theories and knowledge of currents to prevent erosion through superior engineering and designs. In trout streams, erosion and other hydraulic effects produce some of the best holding water for trout, which find great safety and feeding areas under the overhanging cutbanks. Leonardo's studies of vortices in rivers led to his remarkable, ingenious insight that the human heart's aortic valve closes after every heartbeat (to prevent blood going back in the wrong direction), not as a result of muscles, but primarily from the back-pressure of a swirling vortex of the heart's blood inside the neck of the aorta.

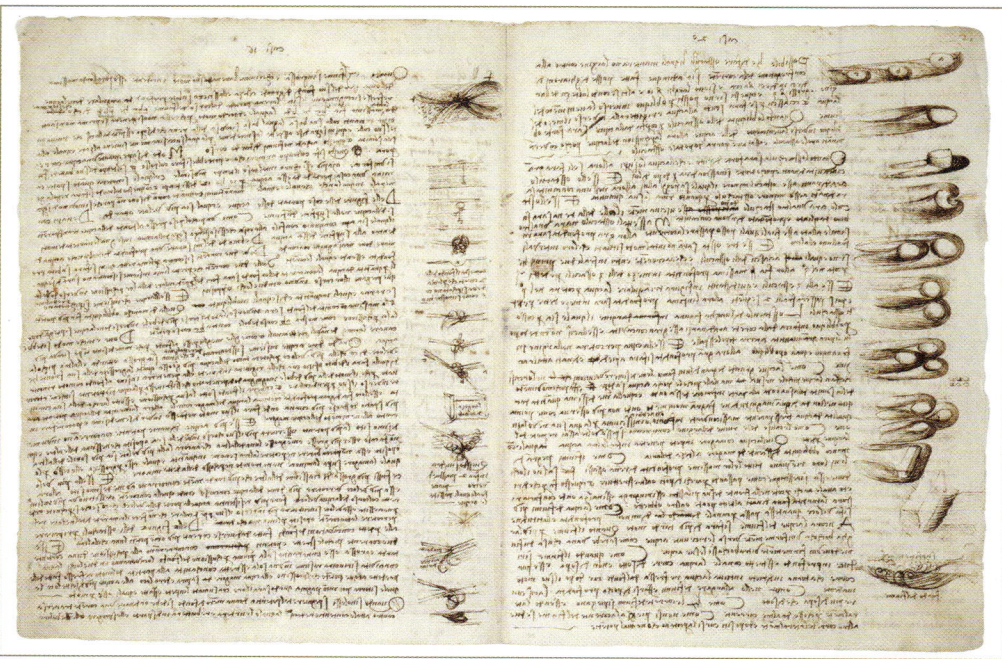

Study of the effect of different obstacles on river currents

This drawing is also remarkable in being virtually cinematic. Although it was an accurate snapshot in time, it revealed the full sequence of what had gone before. The weirs were built; then the waters flowed through; then the vortices began to erode the bank where the weirs directed them. Eventually the erosion came to look like this; and more erosion in the future was easy to imagine.

During his years in Milan, Leonardo spent many days studying and sketching the Adda River. The Adda starts high in the Italian Alps (the Dolomites), near the border with Switzerland, and flows down through Lake Como and onto the flatter lands east of Milan, near Bergamo. While Leonardo was just a boy, the Martesana Canal was designed and built to connect Milan to the

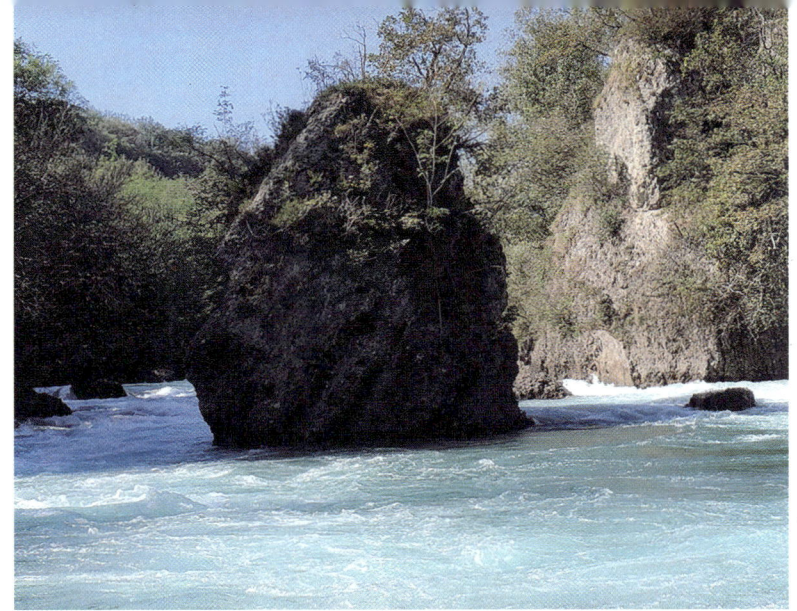

The Tre Corni (Three Horns),
which blocks passage of the Adda River.

Adda, fifteen miles to the east. But a huge rock formation called Tre Corni (Three Horns) and severe rapids made navigation to and beyond the lake impossible. Leonardo developed the eventual solution—starting the Paderno Canal upriver from the Tre Corni—and proposed a technically complex plan both to Il Moro and, later, to France's King Francis I. His recommendation was eventually implemented, but not for another 260 years!

What Leonardo studied and discovered in Milan is as useful to modern anglers as it was to those who paid him for his work. To an angler, "structure" is the term for boulders, trees, and other obstructions in the water as well as depth changes, shelves, rock-strewn bottoms, bends, and points. In the case of difficult rapids and barriers, what is bad for navigation by boat in the main river current can create excellent fish habitat nearby.

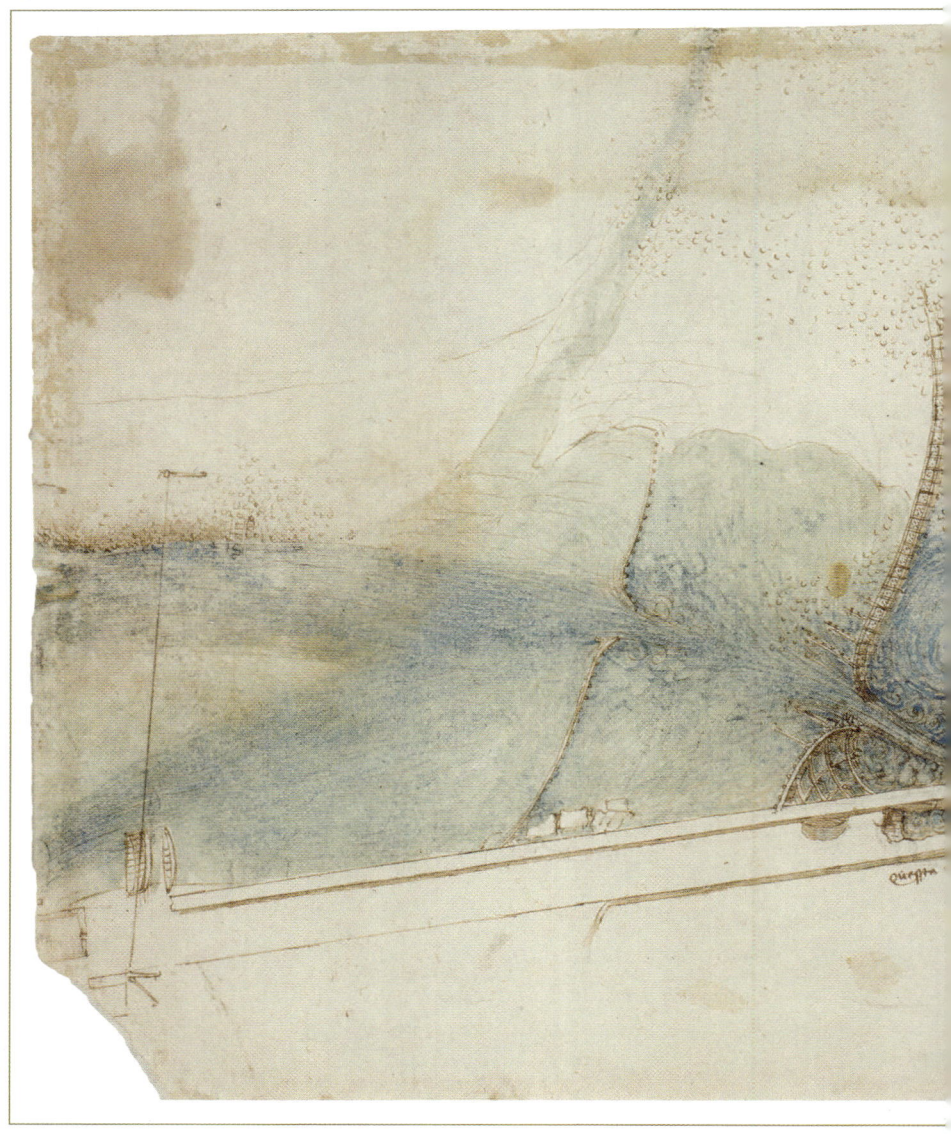

Leonardo, *A Weir on the Arno East of Florence*, 1504

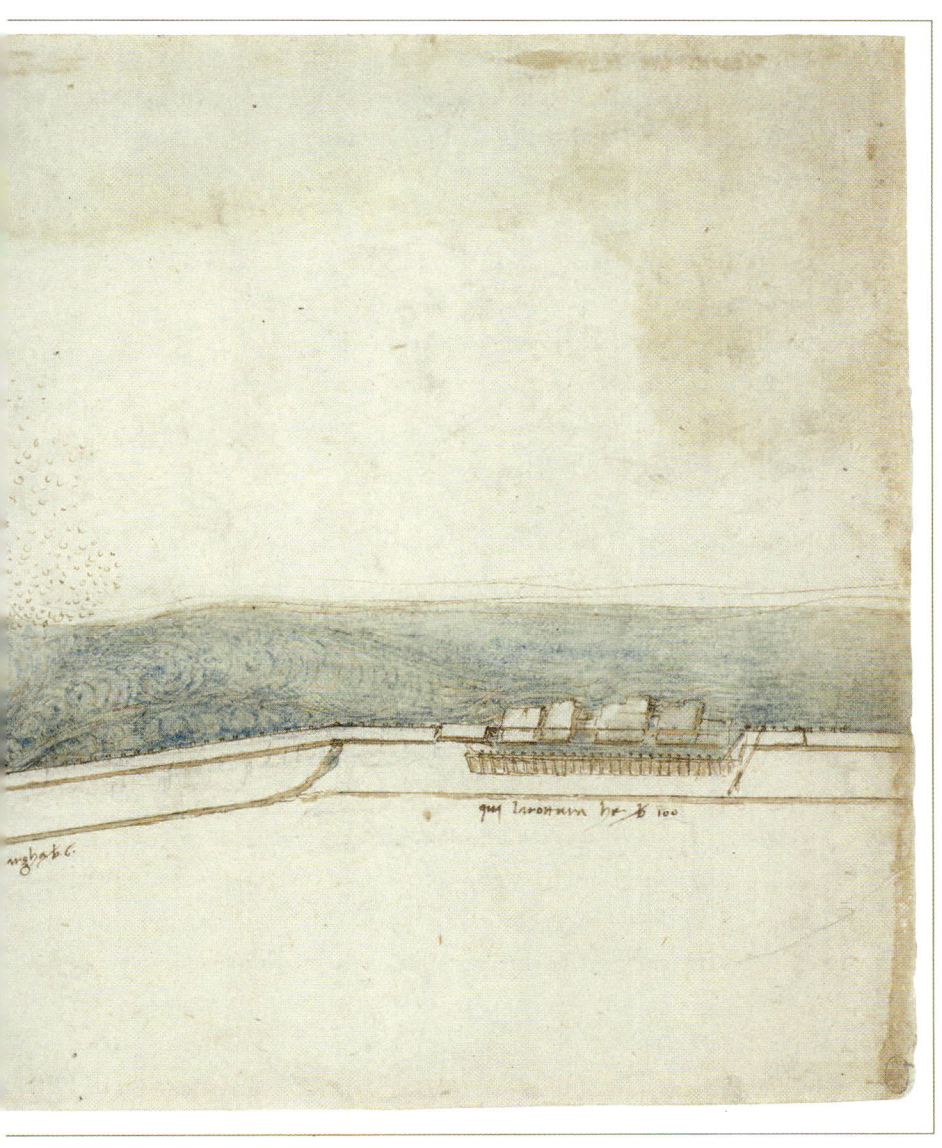

In the Codex Leicester, Leonardo studied the effects of each kind of obstruction in a river (page 44). Before he described how boulders affect currents, which then excavate the riverbed and deposit the excavated sand and gravel, he differentiated various *types* of boulders:

> *Simple round ones; compound many sided; protruding from the water; under the water; vertical; slanting towards the oncoming current; slanting towards the departing current; conical with their bases downwards; conical with their bases facing the sky; and all of these under the water; with concave faces orientated towards the current; with convex faces orientated towards the current; slanting laterally towards the banks.*

These arrangements of rocks or boulders are Leonardo's analyses (and probably experiments) of how currents can be disrupted to control the erosion of riverbanks. The areas he has shaded darkest represent the places where the currents dig deepest into the riverbed around the obstacle. Since those are likely places for trout to hold, the drawings serve the angler as a map of sweet spots to target when we find boulders arranged by nature in similar ways.

Although the sketched boulders appear to stick up out of the water, Leonardo was actually studying obstacles under the river's surface. This makes these illustrations particularly interesting for those anglers who fish with wet flies and nymphs. Seams between fast and slow currents exist out of sight as well as on the surface. Leonardo effectively challenges us to visualize them and the places where they create attractive holding water for us to probe with a fly. Further, if the pictured obstacles are completely submerged, there must be something of a "vertical seam," where faster currents at the surface bring food over the top of the boulder and the pleasantly slower water deeper down behind it. This is likely an important reason that "bumpy water" is so productive for fishing subsurface: a trout does not

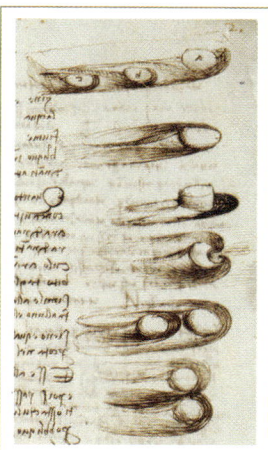

Detail of image (page 44) including a single boulder, second from top

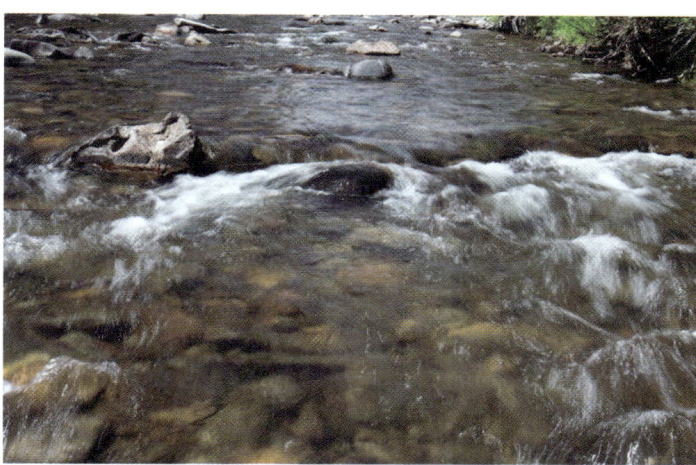

Single boulder pointing upstream, similar to Leonardo's study at left

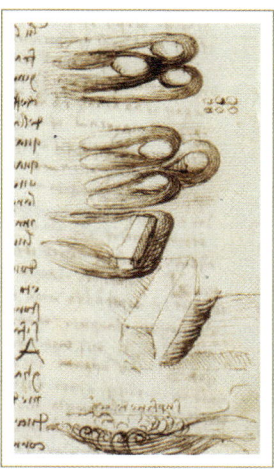

Detail of Leonardo's study (page 44) including a perpendicular boulder in the middle drawing

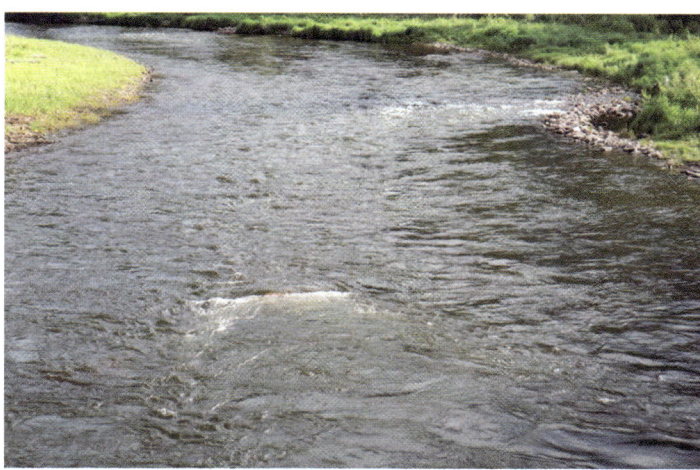

Boulder forming an obstacle perpendicular to the current like the middle drawing at left

need its favorite rock to stick up out of the water as long as it's taller than the fish is high—just a few inches. The trout can hold in the slow current behind a modest rock while a faster current brings food immediately overhead. Leonardo identified this: "Just as a pair of stockings which cover the legs reveal what is hidden beneath them, so the part of the water which lies on the surface reveals the nature of its base, inasmuch as that part of the water which bathes its base, finding there certain protrusions caused by the stones, strikes upon them and leaps up raising with it all the other water which flows above it."

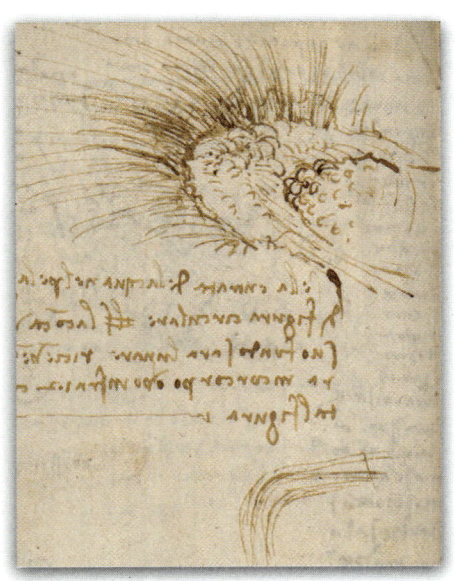

Off the riffle and into the bucket (detail)

Water in the form of a great glacier carved the Conejos Valley in south-central Colorado; the valley has a steep wall on the east side and a great flat floor. The river that is there now, a personal favorite for twenty years, meanders south for twenty-five miles back and forth across that floor. The slightest tilt in the landscape causes the river to shift direction, resulting in many bends and "bend pools" where big trout lie waiting for me.

The water scoops out soil when it picks up a little speed, force, and volume, making even better hiding places for the fish, shaded from the sun and the predacious birds—undercut banks that might extend a foot or two back under the grassy surface—but currents still bring them food. These cutbanks are the mildest, smallest examples of the destructive, erosive power of water.

Across the river from the hill where guide Taylor Streit used to rest his sore back while I fished (somehow shouting corrective instructions to me even when his eyes appeared closed) is a confluence of two parts of the channel that have a bit of everything. Where the two channels of the river come back together after being split by a thirty-yard-long island, the right side has a lovely gravel riffle that pours over a steep shelf into a deep "bucket," like the one Leonardo sketched in the figure shown at left. I stand about three feet upstream from the bucket and happily cast upstream, higher onto the shelf, letting the current carry my flies from the one-foot-deep water of the riffle over the bucket's edge to disappear into the dark green water where the trout lie in wait for stonefly and caddis nymphs to lose their hold on the gravel and wash into their feeding trough.

With gravity pulling the water down the river's gradient, the water speeds up. The water just three feet downstream of me is deep, but I stand cavalierly nearby, confident in my safety because of the excellent footing in three-inch gravel. When I look at the whirlpool, though, I realize that I don't even know how deep it might be. I have cast heavily weighted nymphs into it and never hit bottom. Perhaps it is a hundred feet deep; probably not. But I realize that it is deeper than I am tall (not saying much) and with enough forceful current to drown me, especially with the immobility of water filling up my waders in an instant. So, I move a couple of feet farther away and am completely safe. "He who fears dangers," Leonardo writes, "does not perish by them."

Downstream two hundred yards is a favorite riffle on the left side of the river. It is wide, and I walk out onto the gravel and cast along the shelf, along the seam, and into the slower, deeper water a few feet farther out. As the deeper water comes up onto the gravel, just a foot or so deep, the current speeds up due to the pinch between bottom and surface. The gravel provides perfect footing for me, but the water at my feet is swift now. Within half a minute I feel the gravel at the downstream edges of my boots washing away. Now I'm standing on a new little shelf. Ten more seconds and the gravel under my boots is washing away, and my perfect footing is getting precarious. I will have to move again to avoid danger. This is the dynamic, daily life of a riverbed.

The two major reasons that the waters directly behind and in front of a boulder in the river are so productive for finding trout are the shelter behind it from the force of the main current, and the destructive force of the current on the riverbed around the boulder. The current has dug out the soil and gravel around the boulder to create a deeper place for trout to lie in wait of food and to hide from birds. That same current force digs around the edge of my boots and undercuts my stance on the gravel. It also undercuts the banks to create those "lies" for trout against the edge of the river, which you learn to cast your hopper and other imitations of terrestrial insects into to catch what are often the largest fish in the river. Leonardo documents how those edges slow the current down, which creates better holding water for trout. He also found that there is a "helical" current that transports sediment in a spiral, from the inside curve of river bends back toward the faster central current of the river. Ideally we can visualize that complicated additional flow, pushing our dropper fly toward and away from the bank even as the main current moves it downstream.

Observation never stops,
ca. 1490–95

OBSERVATION AND EXPERIENCE

WHEN I FIRST GET TO A RIVER ON A CRISP MORNING, I'm excited and pretty sure that it's going to be a great day. I'm probably going to catch a world-record brown trout, or at least my best for the month. Fishing is an inherently optimistic sport, like golf, where total failure yesterday does not knock any of the shine off tomorrow's prospects. In fact, failure almost guarantees future success because, after all, now we're due!

Consequently, one of the most difficult things about fly-fishing is to not charge optimistically into the river with rod blazing. Instead, try to act as Leonardo would: stop; observe; think. Stop thirty feet from the river, or at least as far away as you can clearly see both sides of it. Stand behind a tree or sit in a comfortable place in the shade. Spend ten precious minutes studying the river and making mental notes that will add up to finding the fish and catching the fish. Run through a checklist. Guide Justin Spence says, "Let's see what the river gives us today."

If the water is clear, cloudy, or muddy, it influences what color and size of fly I expect the fish to be able to see. If the sky is bright and sunny, or cloudy, or threatening rain (often best of all), it gives me different ideas about where the fish will find security from birds.

I note which direction the sun is shining now and where my shadow will fall when I first step in the stream. The sun is going to move through the day (or at least appear to move, Leonardo might say), and so is the compass direction the river comes from when I walk or wade upstream later in the day. I need to minimize my shadow and the shadow of my moving fly rod and line to avoid spooking fish as that combination of the sun's direction and the river's directions keeps shifting. Sometimes I must cross the river or cast over a different shoulder, taking those shadows with me.

The speed of the river is particularly important. Leonardo was the first to develop the theory of "conservation of flow," observing that the same amount of water passes by every point in a river. I know that where the river gets shallow or deep, narrow or broad, the water will speed up or slow down exactly enough to keep the passing volume unchanged. Trout need that Goldilocks, "just right" combination of food coming downstream steadily, not too much current to fight, and enough depth to avoid becoming bird food themselves. The river may be moving too rapidly or slowly to provide such an ideal spot right here. When I find the right spot and do not see a fish rising to eat on the surface, I have to decide how heavy of a nymph to use to get near the bottom where he should be feeding in that current, and how to avoid dragging the fly in an unnatural way.

Whether I'm fishing among big boulders, normal rocks, or small pebbles will affect where I stand and how close to the fish I think I can get. Are the stream banks hard rock or soft soil, undercut by the current? The fish may be within inches of the bank if there is grass or bushes from which insects can fall in, or they may even be underneath the edge, out of the sun and the sight of birds. If there are boulders or logs, with current-forming cushions in front and slower water and eddies behind them, I need to carefully fish those spots and make my first cast count.

The wind is almost always a factor, both its direction and its speed. For the fly to land where I want it to, I'm going to have to compensate and maybe even cast across my chest or with my unpracticed left hand. A major question from the start and throughout the day is, Where is the trout food? If trout are eating on the surface, it's time to figure out what kind and size of bugs are in the air, emerging or riding on the surface. It's definitely harder if there are no visible insects or fish as clues. Then it's a puzzle to figure out what kinds of nymph-stage insects are most likely to be drifting along underneath the surface and what artificial fly might get a trout to eat. Sometimes it's pure trial, error, and process of elimination.

Most important, I think about what the currents are doing to create places where trout will hold. If structure like boulders or logs are not visibly sticking out of the water, the currents and bumps on the surface can tell us where they are hidden and where the fish are likely to be around them. Ten minutes is a lot of precious time spent not fishing, especially if you see a fish rise in the first minute. But getting in the habit of making simple, careful observations of the type Leonardo lived by, right at the outset, will help you immeasurably to develop a plan for your first hour or two on the river. As a bonus, you will not have blundered into spooking the biggest fish of the day with your first splashy steps into the water. I speak about that mistake from *esperienza*!

None of these are hard, fast, guaranteed-to-work rules; they are Leonardo-logical bets that can and should work out because they improve the angler's odds. Over and over again, Leonardo placed great emphasis on personal experience over received wisdom. "Observation" and "experience" may have been his favorite words; they served as touchstones in his approach to art and science. He criticizes those quoting the wisdom of others rather than their own lived experiences, the ones who "go about puffed up and pompous, dressed decorated with [the fruits], not of their labors, but of those of others."

Or, as Nick responded to something I once quoted, "Sounds like someone who spent more time reading books about fishing than actually fishing."

Neither Leonardo nor Nick were being anti-intellectual here; indeed, Leonardo wrote that "those who are in love with practice without knowledge are like the sailor who gets into a ship without rudder or compass and who never can be certain [where] he is going. Practice must always be founded on sound theory."

Although Leonardo was writing about drawing well, his point is totally applicable to fly-fishing. We need to understand what others have found to be true and said and written. But our own experience must be given the most weight. Fishing should include thinking!

In other words, when what you experience on the river is different from what you read in any fishing book, including this one, you should question the rule. This may be an exception, or it may be something new and important to dial into your thinking in the future. Some people have thirty years of fishing experience, accumulating new knowledge every year. Some people have one year of fishing experience and repeat it thirty times.

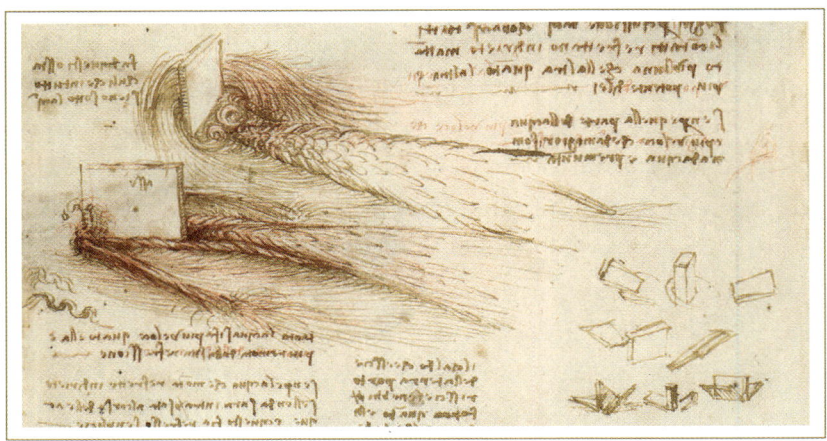

Leonardo's experiments with obstacles
to understand currents in-depth (detail)

I'LL SEE IT WHEN I KNOW IT

For nothing can be either loved or hated unless it is first known.
— Leonardo

MANY OF US USE AN EXPRESSION for something we can't exactly describe, like good art, by saying, "I know it when I see it!"

Leonardo turned this thinking on its head. Drawn to scientific thought, experiment, and experience, he was determined to understand the how and why of things and then express it in his art. His enormous breakthrough in depicting landscapes as they really look to our eyes was the result of his totally original analysis of how light is diffused in air filled with invisible droplets of water. He discovered that distant scenery looked bluer as the day warms up, when the air can hold more moisture. He discovered for himself that the sky and distant landscapes appear bluer because of the sunlight shining through that moisture.

Through intense study of light, geometry, and anatomy, Leonardo determined that light reflects off a curved surface, like a cheek or a nose, in many directions, and that the color of that light as perceived by the human eye changed with each direction. He applied this knowledge in paintings that amazed his teacher, Verrocchio, and then his patrons and the public who saw his work. Once he knew the science, he could see everything differently and skillfully draw it on paper or paint it on a panel. Whether the science was for the painting or the painting was to further the science becomes a chicken-or-egg question.

We too can see more once we understand what is happening in the rivers and how it determines where to fish.

Leonardo wrote and debated among intellectuals that painting was the very highest form of art, the so-called *paragone,* higher in his strong opinion than music or sculpture or poetry. This was not to belittle music or sculpture, which he also excelled in. His argument was based on his firm belief that vision was the greatest of our five senses and that the eye was our most sophisticated organ. Visually studying stream currents and structure is the primary tool for reading the water and finding fish.

Around the time Leonardo began studying water, he obtained access to the written work *Optics and Visual Perception* by the eleventh-century Arab physicist known as Alhazen. Leonardo performed many experiments himself to confirm what he read—a scientific skepticism Alhazen recommended. This earlier polymath advocated the scientific method of hypothesis, experimentation, validation, and revision; the early Renaissance allowed Leonardo to discover this kindred spirit from a different time and culture. They shared a philosophy of not blindly accepting doctrine from any proclaimed authority.

Leonardo,
An Eye in Profile,
ca. 1500

What was this "Renaissance" as it affected Leonardo? When he was barely one year old, in 1453, Constantinople was captured by the Ottomans. Leaders of the Byzantine Orthodox (Greek) Christian Church and other intellectuals fled to Italy, bringing with them a flood of ancient Greek, Roman, and Islamic works written in the Greek and Arabic they understood, and in Latin translations. This Renaissance, or rebirth of classical knowledge, was an enormous boost for humanism. The impact of newly available ancient science and philosophy on the West, especially Italy, was multiplied by the spread of the new movable-type printing press.

Leonardo could not have been born at a better time. Many scholars in Italy had not realized how sophisticated the modern Greeks were. The easterners knew much more Latin than the westerners knew Greek, the original language of the Christian Bible. The Eastern Church had not ignored Aristotle and other pagan writers; they had worked to weave together his logic and science with Christian doctrine. Greek classics were translated into Latin. And the refugees brought with them knowledge from the Arab world, which had not been rejected in the East for being un-Christian but was actively studied for whatever it offered. Alhazen was translated from Arabic into Greek into Latin and, even before Leonardo was born, into the Italian he could read.

Leonardo was never fluent in reading Latin in spite of copying word lists and working at it as an adult. He was hampered by this and said as much in conversation and writing. But another feature of the Renaissance was of great value in helping him partially overcome this handicap: libraries. Wealthy men in Florence and elsewhere had begun to create libraries of their rare, hand-copied manuscripts. With the notable exception of the Dominican monastery of San Marco, where Cosimo Medici donated four hundred books, libraries were not generally open to the public. Leonardo was good at connecting with men of a higher social class who were interested in learning and debating new ideas. His own personality, wit, and charm sometimes got him invited as a guest to exclusive discussion groups that somewhat foreshadowed the salons of eighteenth-century Paris. One or more of these acquaintances likely got him admitted to admire and read works in private libraries, or he sat at San Marco and read a copy—chained to a shelf—of the Italian translation of Alhazen's work on optics.

Leonardo copied references and thoughts from Alhazen's observations and theories into his own notebooks. The subjects were rich: light, reflection, the human eye, and how they affect our direct visual experiences of the sun, moon, and planets as tangible astronomical bodies, not just manifestations of the divine cosmos. Leonardo could extrapolate to his own visual experiences and thoughts about river surfaces. Surfaces were the important boundary between the classical elements of water and air, but they belonged to neither one. Alhazen's encouragement to experiment and prove or disprove hypotheses would have deepened Leonardo's commitment to the prime importance he placed on observation and experience, particularly as superior to received wisdom. He wrote, "Wisdom is the daughter of experience" and, in contrast, "The sorest misfortune is when your views are in advance of your work."

THE EXPERTS

MARTIN KEMP IS THE DEAN OF LIVING LEONARDO SCHOLARS. While he was studying natural sciences at Cambridge in the 1960s, he switched to art history and got involved with Leonardo; he studies him still. Now professor emeritus at Trinity College, Oxford, and past eighty, he remains energetic and overflowing with a broad variety of interests, including art, literature, physics, human behavior, and, especially, perception. It can hardly be coincidental that his lifelong subject, Leonardo, was also intellectually omnivorous and particularly interested in perception.

Kemp returns often in writing, lectures, and conversation to the unfortunate specialization of modern times, which has left many experts in one field virtually unable to communicate with experts in another. He thrives on collaborating with scientists on projects like determining the authenticity of Leonardo's paintings and working for Bill Gates on turning the Codex Leicester into a medium that can make Leonardo's water drawings appear in three dimensions plus time.

Because there are no more than twenty original Leonardo paintings in the world and only one in the United States, it's a very big deal to discover one of the lost ones. If you thought you found one in Aunt Tillie's attic, Kemp is the man you would call. Because of his many scholarly projects, writing, lecturing, and even performing as Leonardo opposite an actress performing as Michelangelo, your call would likely go unanswered, unless you first convinced other leading experts that the case deserved serious consideration. The detective work Kemp and other art historians and technical experts perform—like analysis of the paint, panel, and frame; studying layer upon layer of paint with different wavelengths of light; and researching documents and records that might tie a painting back to primary historical sources—takes years. And for Kemp, these are unpaid years because he refuses any compensation for his expertise, lest his objectivity be questioned.

When we met that first evening in a restaurant three years ago, I asked him what it would be like to have dinner with Leonardo da Vinci. As his book titled *Living with Leonardo: Fifty Years of Sanity and Insanity in the Art World and Beyond* implies, this is not a question that hadn't occurred to him. Kemp explained to me that at first it would be thrilling because Leonardo would tell you so much in such interesting ways. But slowly the tables would turn. Leonardo prepared carefully before a meeting or meal to determine what knowledge you would likely have that was interesting to him and what questions would elicit that information. Eventually he would start asking the questions until, by the end of dinner, you would be sucked dry and left exhausted.

The day following our dinner, I took a cab to the charming village where Kemp lives. He welcomed me into his home and made me tea, and we sat in his living room for hours, talking about the water drawings, Leonardo the man, and what he calls "the Leonardo business." This term includes the fallout from Dan Brown's 2003 novel *The Da Vinci Code*. For years Kemp has received a stream of letters and emails from people he calls loonies. They tell him they have discovered hidden symbols and secret numerology in Leonardo's paintings and drawings, secrets they alone understand. Apparently, these people requesting his help do not take kindly to his polite attempts to dissuade them regarding their discoveries. His website now says that too many inquiries have resulted in too much online abuse from those who did not like his opinions, causing him to stop doing authentication work after forty years. I have heard him accused of authenticating a painting due to a financial motive whereas, on the contrary, he has explicitly eschewed all incentives or compensation for this work from the beginning.

As we sat discussing the water drawings, I became aware that he was sharper at eighty than I had been at forty. He still had everything in a mental rolodex, which amazed me. I'd prepared more than twenty questions for him, and he patiently answered them all. His most surprising response was that he wished we could unknow everything we've been told about Leonardo and just look at the work of the artist, not the work of the great Renaissance genius. Kemp—the man with the most extensive context for Leonardo, developed over decades of research and thought—wants us to look at what is in front of us and ignore what we "know." This is Leonardo's own philosophy of observation in action.

Anglers must be persistent. When fish refuse to cooperate, we must change tactics. First a different fly, then a lighter tippet, a different place to stand, a different depth. Fish make and change the rules; we must adapt.

The British Library turned down my request to study Leonardo's original and important Codex Arundel. I had submitted a formal application with my reasons to view it and a letter of recommendation from a professor who holds an endowed chair at a highly respected American college. I was instructed to study the digital version online (I had) and to read Juliana Barone's catalog for an exhibition featuring water drawings from the codex (I had).

I took the slack out of my line and cast again. Could I impose on Andrea Clarke, the lead curator for medieval and early modern manuscripts, to study, for only a limited amount of time, a very limited number of the codex's pages that were most relevant to Leonardo's investigations of river currents and the nature of rivers on earth? I explained why seeing the originals was important, notwithstanding the extremely high quality of their digital versions. The head of Western heritage for the manuscript collection approved.

On the appointed day I arrived an hour early in order to get my reader card. The British Library is housed in an enormous modern building in London and holds some 14 million books and 160 million additional printed and other objects. It and the Library of Congress are the two largest libraries in the world. Since at least 1610, a copy of every book, newspaper, recording, or anything else published or distributed in the United Kingdom or Ireland is required to be "deposited" at the library. None of these may be checked out or removed from the building; the British Library exists to preserve and collect for research only. Each year the library adds six miles of shelving just to hold new acquisitions. Since 2013 it has also collected every digital page on every .uk website, one billion pages and counting.

I had come for something at the opposite end of the publishing spectrum: the one and only copy of an unpublished manuscript, all in Leonardo's handwriting and drawing. On the first floor are clear signs leading to the ordinary computer-filled room every first-time researcher must enter for admission anywhere in the library. Once I had filled out my forms on the computer, I was called over to the first available clerk's desk so she could ask me questions and issue me a reader card. She saw that I was heading to the manuscript room and asked me what I was going there to see. I told her the Codex Arundel, and she exploded, "The Codex Arundel? Nobody gets to see the Codex Arundel!"

"Hey, Betty," she shouted across the room to another clerk, "he's going to see the Codex Arundel!"

"Wow!" Betty said. "You are so, so lucky! Everyone wants to see that. What are you doing that they let you see it?"

Thus ensued a happy, spirited discussion among the three of us about what I was trying to research and write. They enthusiastically wished me good luck, and I walked out of their office full of gratitude, pride, and good cheer.

On the second floor, I walked past a multistory glass enclosure of beautifully bound books and opened a door marked "Manuscript Reading Room." It was a totally hushed, well-lit room where a dozen people were intently studying some treasure in their particular field. I showed my pass to a special security guard just inside the door, and he pointed me to a typical-looking library counter. I stated that I had an appointment with Kathleen Doyle, and she told me to wait.

Soon a pleasant woman in a Covid mask came to greet me, and I followed her to a small, private room. To my surprise, Doyle, the lead curator for illuminated manuscripts, is an American. We discussed how she came to be in charge of medieval manuscripts, illustrated and often hand-colored, at this elite British institution. It turns out that she was a high-powered lawyer with a Seattle law firm that sent her to work at their London office some years ago. After a while, Doyle decided to give up practice as an attorney and follow her academic passion for books written by hand, before the advent of the printing press with movable type around 1455, and illuminated—illustrated—by hand with highly detailed miniature paintings, often by extraordinarily talented artists. These include small prayer books for nobles who could afford the luxury, monastic books and papers, and other works of learning on many subjects from before Leonardo's era.

Doyle had agreed to oversee my visit while Clarke completed final preparations for opening a blockbuster exhibition, *Royal Cousins, Rival Queens*, that she had worked on for years. It covered the complex and eventually deadly relationship between the Protestant Queen Elizabeth I of England and the Catholic Queen Mary of Scotland, telling the story with precious portraits, jewelry, and manuscripts from the British Library's collection and others.

With this caliber of experts, I felt like Einstein was substituting for Shakespeare to sit and hold my hand during the visit.

Doyle unlocked a wooden box and took out one of the true jewels of the British Library and set it on a wooden stand for me to study. Leonardo did not write the Codex Arundel as a single notebook. All of its 283 handwritten leaves, each with a front and back page, had been purchased by an early and great connoisseur, Thomas Howard, the fourteenth and twenty-first Earl of Arundel (don't ask), and bound into a single volume for his interest and enjoyment. His grandson, Henry Howard, the sixth Duke of Norfolk and twenty-fourth Earl of Arundel, was persuaded by John Evelyn to donate the collection to the Royal Society in 1667, which at the time was formed by twelve men, including Christopher Wren, Robert Boyle, and Evelyn. In 1831 the British Museum purchased the Leonardo Codex Arundel and 549 other manuscripts from the Royal Society. This huge treasure was just half of the original Arundel collection! When the British Library was created in 1973 as an offshoot of the museum, Leonardo's manuscript was part of the Arundel manuscripts transfer.

Doyle carefully went through the now-unbound and separately protected pages of the manuscript to the first of the key pages I had asked to view. She sat nearby, typing her own work into her laptop, while I took in Leonardo's Tuscan "mirror" handwriting (see pages 86–87) and studied the illustrations he had sketched in the small margins. These are not the elegant drawings held at Windsor Castle but rather his quick renderings of currents and rivers illustrating the hydrologic points he was making in adjacent words. Leonardo thought spatially as well as verbally. Neither would be adequate by itself to convey his concepts and understanding; these were his thoughts laid out in both forms for completeness.

The lines Leonardo drew on the page are sparse but somehow fully express how rivers work. Of particular interest and importance, he drew how the

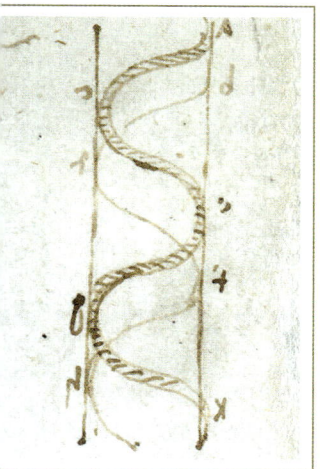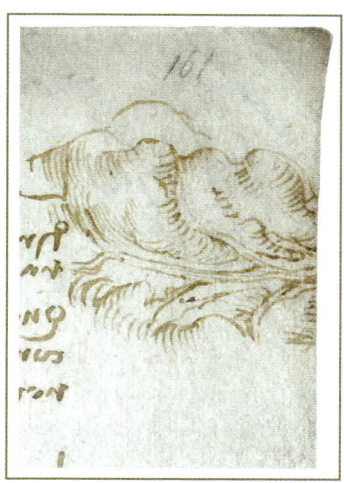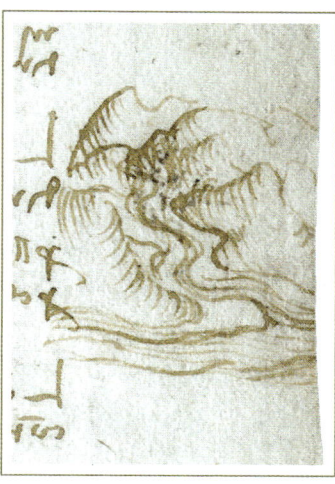

In the detail at left, Leonardo illustrates how the main current of a river reflects off alternate banks. In a thinner, lighter shade of ink, he portrays where the main current will be in the future as the present current erodes the river's banks and bed. In the details at middle and right, he depicts rivers making and destroying mountains in simple, deft, charming sketches.

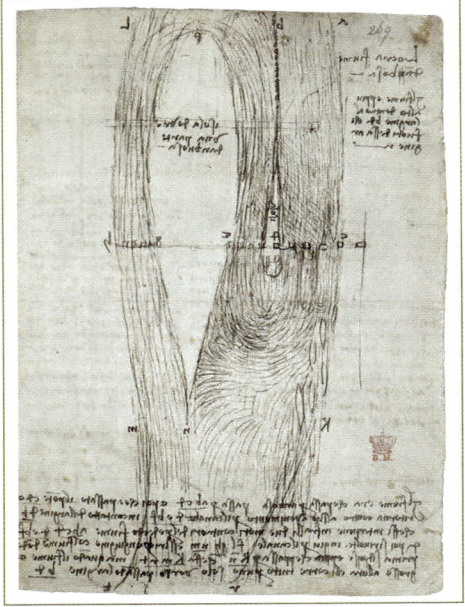

The large, teardrop-shaped island that splits the Loire River was part of the French town of Amboise, where Leonardo lived from 1517 to 1519. The unusual currents shown in this drawing from the Codex Arundel created hydraulic problems for Leonardo to solve and would present several excellent fast-water/slow-water seams for anglers to explore.

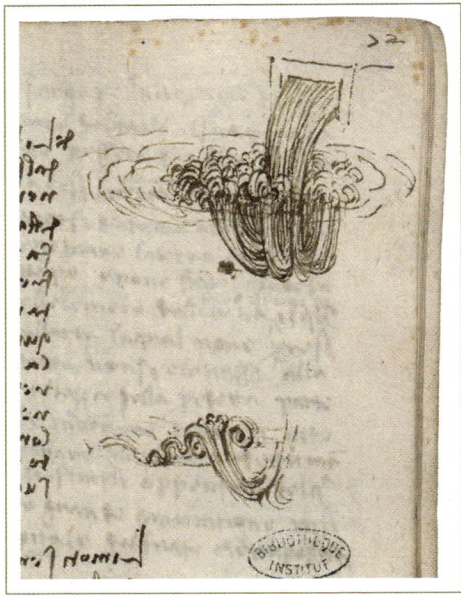

Leonardo was fascinated with the manner in which water "bounced" off river bottoms and off water, when it was filled with air bubbles; the resulting dissolved oxygen makes the surrounding water more attractive for trout (detail).

layer of water he thought floated between the earth's center and its surface came out in mountain springs that made their way as rivers down the valleys they eroded, toward the sea. This was the circulation system of the "macrocosm" (earth, the greater world), which he believed at the time was exactly analogous to the arteries and veins of the "microcosm" (the human body, the lesser world).

Each time I finished studying and making my own sketches and descriptions of what I saw (no photography is allowed), Doyle would interrupt her work to turn the leaves to the next page I had requested in advance. She generously offered to get me high-resolution images from the library's photographer if I requested them.

When I was done, she carefully closed the leather binding of the protective covers and placed the codex back into the wooden box, locked it, and checked that the box was chained and locked to the cart that bore it into the room and back out again. We talked a bit about life in England and the wonderful opportunities for study and travel the country offers; then my time was up. Again, excited by the special period I had spent in the direct presence of Leonardo's work, I thanked her, the desk manager, and the security guard, left the inner sanctum, and went back out into the normal world.

THE ULTIMATE THANK-YOU GIFT

THEY CHOPPED OFF HIS HEAD. Kings died in battle or were assassinated by rivals, but for a parliamentary government to put a crowned king on trial and then execute him—a king who was believed to be chosen by God to rule—was something new under the sun. Every nobleman whose property and title had been granted by a monarch and guaranteed to their descendants by the divine right of kings realized that a political earthquake had destroyed the stability of a system older than memory.

So it was that Charles I, King of England, Scotland, and Ireland, was beheaded outside London's Whitehall Palace in January 1649, by order of a court of sixty-seven commissioners chosen by Parliament, following a trial accusing him of being a tyrant. His son Charles II continued to fight the forces of Oliver Cromwell for over two more years, until the final military defeat of the king's forces ended the English Civil Wars. Charles II, at one point hiding in an oak tree only feet away from soldiers seeking to capture him, managed a storied and very narrow escape from England to France, then to the Netherlands and Spain.

One of the nobles who stayed loyal to the deposed king, and fled England for Antwerp at the beginning of the Civil Wars in 1642, was Thomas Howard. Howard's ancestors had been executed and stripped of their lands and titles seventy years earlier by Queen Elizabeth I for participating in plots to put the Catholic Mary, Queen of Scots, on the Protestant English throne. Howard happened to be Europe's greatest art collector and owned a bound collection of more than five hundred drawings by Leonardo, purchased from the heirs of Pompeo Leoni, a sculptor from Corsica.

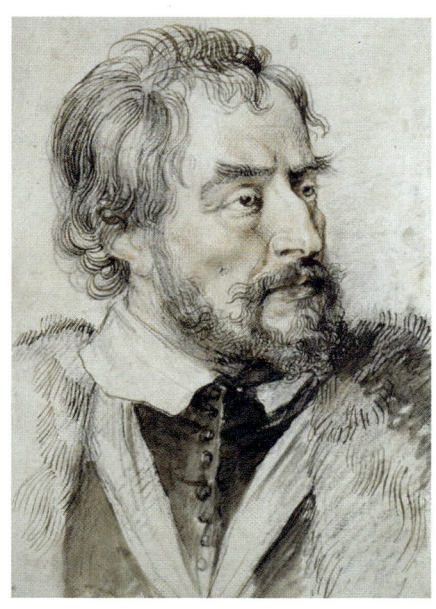

Peter Paul Rubens,
Thomas Howard, Earl of Arundel,
ca. 1629–30

After Cromwell himself became a military dictator and then died in 1658, Charles II was invited in 1660 to return from his exile in Europe and assume the throne, which he did. Howard's grandsons, Thomas and then Henry, back in possession of lands and earldom, had the premier noble title of Duke of Norfolk restored to them by the king at the request of Parliament. Henry Howard probably gave the entire book of Leonardo's drawings to Charles II as his way of saying thank you. Although the gift made Charles II (an angler, by the way) the fifth owner after Leonardo in just 150 years, this treasure stayed together for the next 365 years and remained in the British royal family's collection, today held in trust by King Charles III.

IN SEARCH OF THE HOLY GRAIL

MARTIN CLAYTON IS A SERIOUS MAN WITH A SERIOUS JOB. Officially, his title at the Royal Collection Trust is head of drawings and prints. Like the English, the title is understated. I would call him the "guardian of Western civilization and culture." He was appointed, however, not by an ebullient American, but by the trust that oversees the British Royal Family's unparalleled collection of castles, weapons, furniture, clocks, and artwork. There are more than a hundred thousand drawings and prints acquired since 1641 by the Houses of Stuart, Hanover, and Windsor, notably including 550 drawings by Leonardo da Vinci, the ones from that nice thank-you gift.

Clayton entered the small and businesslike security office / reception room, down a few stairs, at exactly the agreed-upon time. There is nothing regal about this setting. But I had already checked into the official pass office outside the thick castle walls and then approached, uphill and imposing, the Henry VIII Gate of Windsor Castle. Three men in uniform bearing machine guns were friendly in their greeting at the gate, as well-armed and -trained men can afford to be. They told me where I needed to go, across a wide stretch of castle, past perfectly manicured lawns, past the enormous and stout tower that protected medieval kings and their courtiers, through a small archway, to a modest doorway marked "Side Door Private." I stopped along the way to watch a group of Queen's Guards, in their scarlet coats and tall bearskin hats, march in perfect formation up the pavement to their posts. The castle was closed to the public on that cold, sunny, blustery day; I was alone and chose to imagine that this was all for me. And it was regal, indeed.

Clayton led me up a flight of stairs and into the print room, which is actually three large rooms surrounded by bookcases. They are not tall rooms, and the ceilings are made of embossed metal panels that Americans consider antique but are not old by castle standards. The rooms were established in the 1850s by Queen Victoria and Prince Albert, who spent many evenings enjoying their vast collection. Instead of books in the bookcases, there stand side by side hundreds of hinged boxes covered in beautiful red leather and stamped in small gold letters, RCT; each one contains treasures. For Clayton, this must have seemed like a normal day at the office, and I was only one of his many responsibilities that day. For me, it was the entry into Heaven.

When I had written six months earlier to request an appointment, I doubted I would receive a response, much less an approval. I had been studying these drawings and all things Leonardo for more than two years, but I had no academic credentials nor expectations of being taken seriously as a researcher. I wanted to closely study the drawings firsthand, not only in books, but who wouldn't? To my great surprise, though, I did receive a response from Clayton—clear, polite, and encouraging. Due to Covid, no one had been admitted for many months, he said, but write back near the fall and we would work out a date.

I went into furious planning mode. List the key drawings to see and why. Try to arrange meetings with additional legendary authorities on Leonardo to ask questions only they could answer. Try to see Leonardo's Codex Arundel at the British Library, filled with sketches of river currents; make a schedule

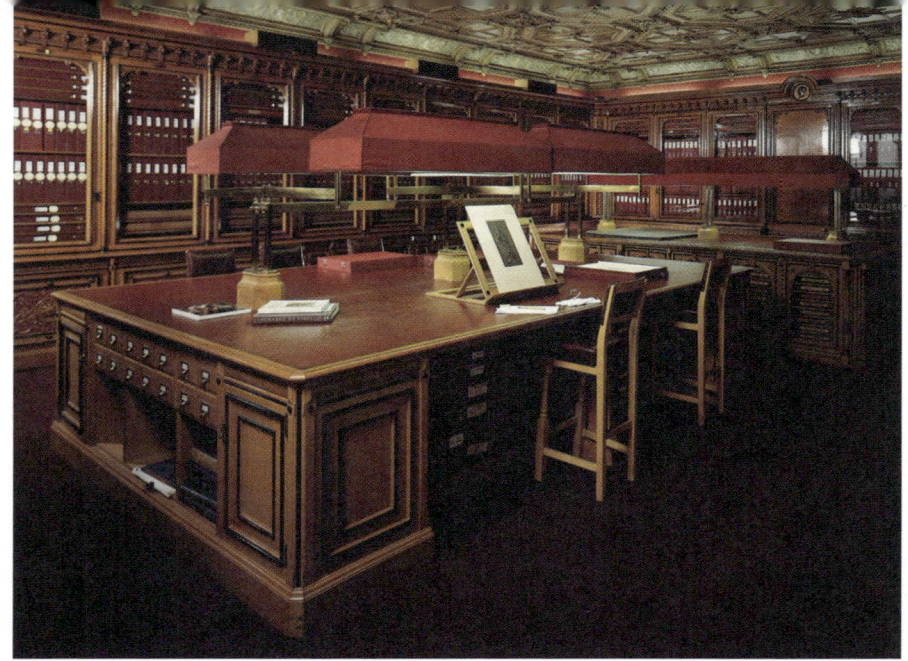

Royal Collection Print Room, Windsor Castle, London

to visit the huge and famous *Burlington Cartoon* in the National Gallery in London, in which Leonardo depicted the Virgin Mary's feet immersed in water; see Verrocchio's painting nearby of *Archangel St. Raphael with Tobias*, because Leonardo painted not just the translucent dog but the *fish* whose gall will cure Tobias's father's blindness. Schedule Covid tests coming and going; fill out forms that conflict with each other, and study official websites that change every day with new and confusing regulations. Even with everything set, I was unsure I would get into England and, later, unsure I would get out again.

Finally, all of that was behind me. The previous evening had been spent timing how long it takes to walk from my hotel to the pass office and finding the right gate to enter Windsor Castle. And reviewing what might go wrong that would prevent my using every moment of the precious time I'd been allotted. At last, I was surrounded by tranquility and filled with excitement.

Clayton suggested that I spend a couple of hours studying drawings on my list and then ask him the questions I had prepared. He set me up to work at a large, well-lit surface that covered much of the room, together with an easel, a magnifying glass, and a tall stool, and invited me to give him the number of the first drawing I wanted to see. I asked for RCIN 912660, the 1510–12 drawing Leonardo made of water flowing through a square hole in a dam, forming a waterfall. I had kept a copy of it on my desk for many months, fascinated by its unprecedented depiction of water both above and below the surface, the foamy air bubbles, turbulent currents, and eddies. It also reminded me of one of the most beautiful pools I've ever fished, a long and difficult hike up into an intimidating canyon just a few miles from my cabin. The connection between the fly-fisher's thinking and Leonardo's thinking is perfectly obvious in this work.

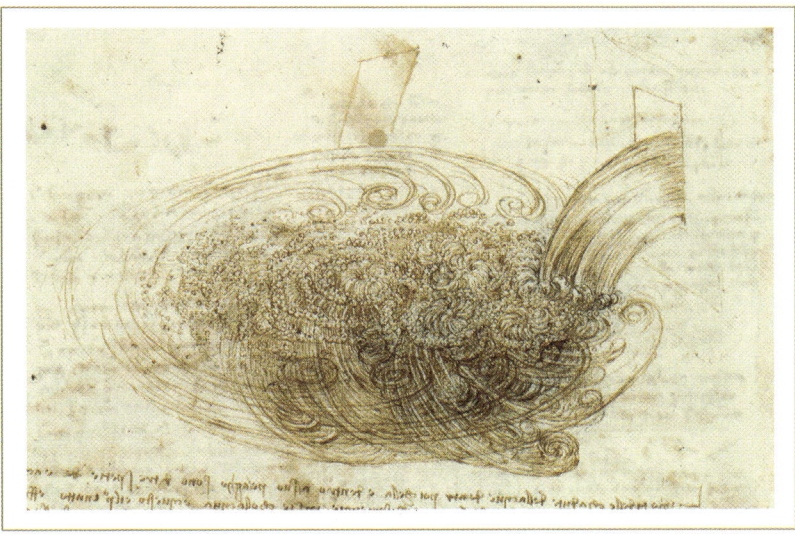

This unprecedented drawing illustrates Leonardo's understanding
of subsurface currents and turbulence (detail).

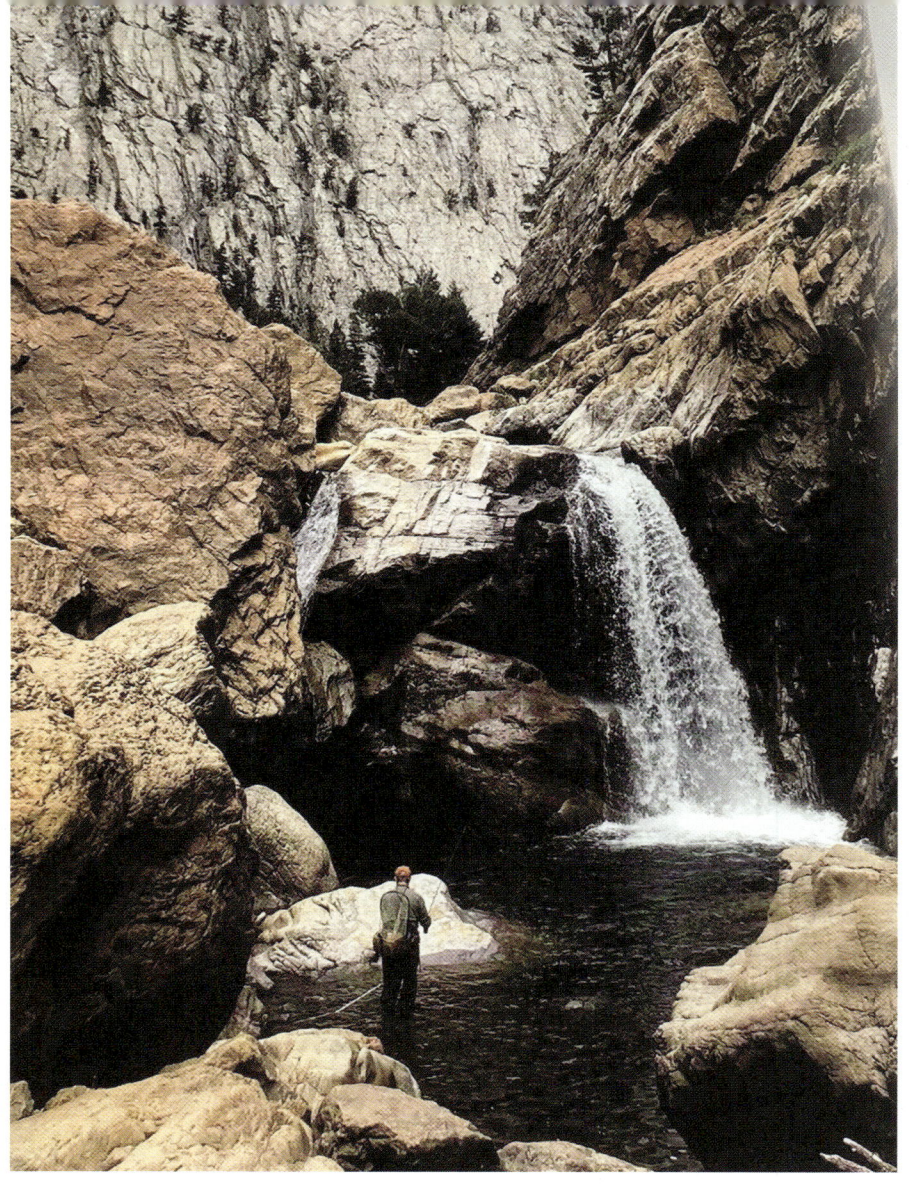

The author fishing what we call Leonardo Falls,
in a 2,000-foot-deep box canyon

Clayton studied his index book to determine the right box number, gently hefted the large box onto the table, opened its clasps, and turned back the lid to reveal a group of carefully stacked, loose mattes, each enclosed in a thin, hard layer of material that protects the drawing from ultraviolet and other light that can fade the ink and destroy the paper over time. For the sake of their preservation, he demonstrated exactly how I should look for and handle the desired drawings, put the first one up on the small easel for me, and left me to my work. I tried to be calm, but in my head, music swelled and choirs sang; it was a purely wonderful moment.

Clayton went back to his desk and worked at his computer. Although he was tending to me as a librarian, I was very aware that he is an extraordinary expert. I had read his books on Leonardo's drawings and on the breadth and history of the Royal Collection's works on paper. The foreword to one of his books is written by the present King Charles III. I had watched him interviewed on the BBC and knew he had organized exhibitions bringing Leonardo to museums and the public across the United Kingdom and around the world. Restorers working to preserve drawings and assistants preparing loans to other institutions came in and out of the room to report to him and consult with him. I was the one researcher admitted for the day and took up a lot of his attention; he never complained.

Sometimes he let me go through a box to find the relevant drawing; with the most important ones, he preferred to handle them himself, from the box to the easel before me. I took my notes without rushing and, after assigning all copyrights to any nonflash photographs, took lots and lots of photos of particularly interesting details of each drawing. Although every drawing was memorably unique, I experienced three overarching surprises from taking them in as a group.

The first was that Leonardo's drawings are small. In all the books I had studied, they appear monumental; each could well be two feet on a side—complete works of art. In person, they are measured in inches, in many cases the size of an index card. Second, they are amazing, beautiful. Even in the sketches, the details are sure—no hesitation and little reworking of lines. We know Leonardo as a great artist from paintings he spent years working on, but these drawings and sketches in ink (pencils did not yet exist) have shading and depth to them and tiny pen marks that are easy to interpret, especially with a magnifying glass. They are precise: no doubt that the meandering rivers he mapped actually bend exactly as he drew them. Finally, they are full of power. What looks interesting or appealing in a book is on close inspection deeply fascinating, charming, or even intimidating.

After two hours studying individual drawings and returning them to their boxes and Clayton bringing me more, it was time for my questions. I asked him about Leonardo the man as well as the artist. What would he like to see corrected in the popular misunderstanding, and what would he like people to understand? Clayton told me that the portrayal of Leonardo as a stiff man speaking as an oracle is just wrong. He was warm and human enough to be good company, to plan and execute entertainments for the duke of Milan's court, and to be the ornament of the French king's court in the artist's old age. I asked him about Leonardo being the model for Verrocchio's *David* sculpture, and Leonardo's student Salai being the model for Leonardo's many drawings of a beautiful young man's face. He shook his head and explained that these full-lipped, curly-haired boys are archetypes that Leonardo used before he ever met Salai. Leonardo-as-model is just a version of a fantasy dating back to Plato, that a beautiful mind should reside in a beautiful body; he said that we simply do not know what the actual young Leonardo looked like.

In the final two hours, I worked unhurriedly through the rest of my list—more than forty water drawings in all, plus three beautiful sketches of eyes. Leonardo not only believed eyes were the window to the soul; he was also deeply interested in optics and how the eye and brain worked together to perceive the world.

I did a double take when I looked closely at a drawing of the Adda near Villa Melzi through a magnifying glass (next page). Could these men be fishing? I asked Clayton what the man on the path was carrying. He said some sort of tool. But immediately seeing right through me, he asked, "Do you want it to be a fishing rod?" Not "Do you think it's a fishing rod?" but, implicitly, "What sort of wishful thinking are you indulging today?" I laughed and said, "Exactly. I really do want it to be two fly-fishermen."

I asked how often people come to study these amazing Leonardo drawings of rivers, currents, whirlpools, and deluges. He thought about that and told me the last person to do so was Tulane professor Leslie Geddes, three or four years before, while she was researching her excellent book *Watermarks: Leonardo da Vinci and the Mystery of Nature.* Although the Royal Collection admittedly includes everything *From Holbein to Hockney,* as one of Clayton's books is titled, it still felt to me that the water drawings are underappreciated.

Before I left for the day, Clayton generously offered to show me a few of the spectacular large anatomy drawings Leonardo made around 1508. The sureness of his lines depicting muscles, bones, and blood vessels, drawn with a goose or swan quill pen and ink in the steadiest of hands, struck me as near impossible for a human. But there they were, illustrating truly new knowledge for mankind. Leonardo depicted details never understood before and which confirmed his belief in the perfection of God's natural creations, with nothing excess and nothing missing.

I shared with Clayton that, in spite of having seen each of these Royal Collection drawings countless times in books, I was totally unprepared for their beauty, their small size, and their overwhelming power when seen in person. The man who spends his days surrounded by priceless original works of art smiled and said, "Reproductions just don't do it."

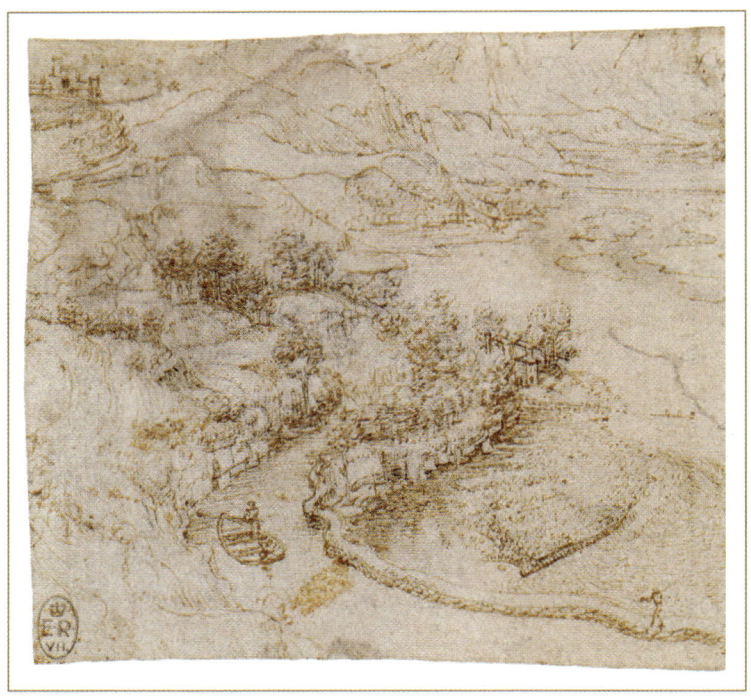

This drawing was made while Leonardo spent the year 1512
at Villa Melzi on the Adda River (detail). Near the middle of the image
are rapids; in the foreground, the left bank is natural hillside and the
right bank is stabilized with stones. A faster current seems to be
entering the river from across the path below the boat.

At 4:30, precisely at the end of my allotted day, I packed up, thanked my host profusely, and took my leave. I hardly remember walking back to the hotel. I ate dinner alone at a fifteenth-century place serving old English recipes. Although the food was very good, I spent the evening slowly shaking my head back and forth, incredulous at what I had seen, felt, and experienced. I don't think I'll have another day like that one as long as I live.

A VERY CURIOUS MAN

The greatest gifts are often seen to rain down in a natural way on human bodies through celestial influences, and sometimes supernaturally, combining beauty, grace, and merit fulsomely in a single body. These gifts are made apparent whenever such a man turns his attention to something. His action is so divine that he leaves in his wake all other men, making it clearly evident that this quality arises (as it does) from the gift of God and is not cultivated by human artifice. Men saw this in Leonardo da Vinci, in whom there was not only beauty of body—never praised enough—but also a more than infinite grace in all his actions. Such was his resulting virtue that whenever his mind turned to difficult matters, he wholly resolved them with ease. The strength within him was conjoined with dexterity, and his soul and his courage were always regal and magnanimous. And the fame of his name grew so widely that it was not only held in esteem in his own time but also increased in posterity after his death.

— Giorgio Vasari, *The Life of Leonardo da Vinci*

SO BEGINS THE FIRST BIOGRAPHY OF LEONARDO, published in 1550. Vasari's praise is over the top for any mortal being, and eventually he sprinkles some criticism into his description of Leonardo's life, which had ended thirty years before. While Vasari tactfully avoids discussion of his subject's sexual orientation, he does assert that Leonardo was a procrastinator and a heretic, failed to finish his projects, and wasted his artistic talent by spending so much time on science and other relatively unimportant pursuits. Nonetheless, for those who never met Leonardo (as Vasari had not), this biography became the baseline of fact and legend until modern historians began to research original documents, particularly Leonardo's notebooks, at the end of the nineteenth century. I have only two points to make here. First, that Leonardo was fabulous, but not perfect, as this often-repeated introductory description makes him out to be. And second, that Vasari does Leonardo an inadvertent disservice by stating that solutions to difficult problems came easily to him, as gifts from God. I believe that much of Leonardo's genius came from constant hard work. He would have appreciated Thomas Edison's remark that "genius is 1 percent inspiration and 99 percent perspiration."

So what was the truth about Leonardo's foibles?

As he walked about town, handsome and tall in his short green coat and pink hose, Leonardo was both a flashy dresser and a bit out of fashion, since men his age were wearing longer garments. But his outward appearance was only one minor feature of his multifaceted and unusual personality and character. He sought to rise in the society he inhabited in spite of the handicaps it placed on him. He managed to be ambitious for fame and fortune within conventional standards and yet quietly maintain his own, quite different, personal standards of success. He sought financial security and also wrote critically about those who focused primarily on material comforts. He sought artistic commissions but then frustrated his patrons by not completing them, earning a mixed reputation through unforced errors. He believed in God and worked for, and in, churches, yet did not spend much time worshipping nor truly accepting some of the fundamental precepts of church doctrine. What he sought the most—a comprehensive and unifying understanding of all the world, large and small—required more of his time, attention, and difficult mental work than others thought valuable. Even if he achieved those ends, they would not immediately provide financial security or status. These pursuits were to satisfy his own boundless curiosity. He envisioned sharing his discoveries with the world through books, but he seemed in no rush to publish what he did learn, possibly because there was always so much more he wanted to know. He kept most of what he learned to himself, even though printing with movable type offered him the opportunity to share his many insights and discoveries with the world. While the judgments of others were important to many aspects of his life, ultimately Leonardo would judge himself.

Procrastinator?

Leonardo certainly took his time with commissions, left many unfinished, and even got sued by a frustrated patron. When he was well past the expected completion time in 1498 of *The Last Supper* in Milan, the prior of the convent that had commissioned the work asked the duke of Milan (who was paying) to intervene. Leonardo was called to account to the duke for his extreme tardiness. He defended himself at length, claiming that great artists do their most important work thinking before applying the brushwork. He explained that he had not yet found a face suitably beautiful to represent Christ, nor one suitably ugly and wicked to represent the treacherous Judas Iscariot, who was the villain of Christ's shocking statement to the assembled apostles in the painting, "One of you will betray me." Leonardo told the duke that if he was required to finish the work immediately, he could always use the face of the prior himself to represent Judas. Vasari reports that Il Moro burst out laughing at this outrageous riposte and agreed that Leonardo must be granted the time to perform his task as the artist thought best.

Anglers should, by analogy, do their important planning and thinking *before* they cast.

The *Mona Lisa* portrait is another example of perfectionism perceived as procrastination. Leonardo was commissioned in 1503 by her wealthy merchant husband, Francesco del Gioconda, to paint her portrait to celebrate their new child and new home in Florence, where Leonardo was living again. For ten years Leonardo continued to work on the painting, taking it with him as he moved, from Florence to Milan, Rome, and ultimately to France, never delivering it to Sr. Gioconda. Everyone who saw it was amazed at its extraordinary lifelikeness, but Leonardo continued to work on making it more perfect. It may have been finished around 1513, or never finished to Leonardo's complete satisfaction at all. He kept it with him to the last year of his life.

Leonardo was also busy with his own work. He continued to study and analyze human anatomy, the currents of rivers, the nature of light and shadow, geometry problems, and the flight of birds, which he hoped with good reason would lead to human-powered flight. His search for knowledge of natural effects and their underlying causes kept him busy observing, sketching, and writing. Vasari blames him for wasting his precious time and talent in such pursuits, but Vasari's interests lay in art and architecture, not in the countless subjects that fascinated Leonardo. The man could never be accused of laziness.

A business partner of mine once told me that fishing is a waste of time, especially catch-and-release fishing, which doesn't even put food on the table. No fly-fisher who spends hours hiking to their favorite spots, looking for rising fish, changing flies over and over in pursuit of perfection, untying maddening tangles in their line, studying the patterns of river currents, and enjoying being out in nature would ever agree to such a silly proposition. We should respond to judgmental statements like the one Vasari and my partner made with, "Wasteful to whom?"

Secretive?

Some contemporaries claimed that Leonardo was a man who must be up to no good because he wrote in code or because he avoided sharing the vast majority of his many notebooks with others. They may have thought that he practiced occult rites like alchemy (which he actually detested). He certainly did not widely discuss his dissection of corpses, which was illegal, in spite of his producing the most useful knowledge of anatomy and deductions about physiology since Roman times thirteen centuries before.

It is true that Leonardo was secretive about two things. First, he was not open about his work on the project of flying. He believed that the man who discovered how to fly would reap everlasting fame. His tireless, meticulous investigation and documentation of precisely how birds fly, how their wing muscles work against gravity, and how they use their wing feathers and tail feathers together to work with the winds to achieve speed, direction, and lift were all highly proprietary. Four centuries later the Wright brothers practiced similar secrecy with their many experiments in pursuit of the same goal. And they achieved the respect and lasting fame Leonardo predicted.

Fly-fishers also keep a few secrets. They are usually generous with advice on where the fish are rising, whether they are feeding in fast or slow water today, and what kind of flies they are taking. But perhaps we don't reveal exactly where the largest fish we missed is still hiding or a spot that doesn't look very fishy but actually is.

Second, Leonardo kept his invention of an underwater breathing device a secret. It held air in a wineskin and included goggles for the diver to wear over his eyes. He wrote: "How and why is it that I do not describe my method for remaining underwater and how long I can remain there without coming up for air? I do not wish to divulge or publish this because of the evil nature of men, who might use it for murder on the sea bed." He believed that it could be devastating in war, with divers drilling holes in the bottoms of ships and sending countless sailors to their deaths. He railed against the inhumanity and futility of war but designed instruments of warfare anyway in order to impress and support his patrons engaged in battles. And he was especially concerned that his invention might be used by hostile forces against the army he was working for.

Leonardo's backward handwriting was believed by some to be a code for a man with secrets to hide. It was not secret code at all. Since he was cursed (in the general opinion of his time) to be left-handed, he wrote backward, from right to left, to avoid smearing the ink of what he had just noted. When writing for others to read, he did so in the normal direction so that they could understand what he presented on maps and proposals, but it would have been slower for him. Here is a single example of the mirror writing, so you can appreciate how not-secret it was:

A sample of Leonardo's "mirror writing"

Anyone looking at this would immediately see that the capital D's are written backward. They could likely suss out some of the other letters in Leonardo's handwriting to then know that the whole thing is written backward.

The image above reversed, as if in a mirror

Holding the page up to a mirror and looking in that mirror would make the handwriting read conventionally, left to right. So, if you could read his handwriting, you would have little trouble transcribing the message into legible Tuscan-dialect Italian: "Del modo del notare de' pesci; del modo del lor saltare fori delle acque, come far si vede a delfini che par cosa maravigliosa formare salto sopra la cosa che non aspetta, anzi si fugge; Del notare delli animali di lunga figura, come anguille e simili; Del modo del notar contro alle corenti e gran cadute de' fiumi." The hardest part is probably being fluent enough in Tuscan, Italian, and English to do a good translation, like this one by Domenica Laurenza and Martin Kemp. The brief items in the list are various subjects Leonardo is telling himself to explain to his future readers—and there are thousands of them:

> *On the manner of swimming of fishes. On the manner of their leaping out of the waters, as the dolphins are seen to do; and it seems a marvelous thing to make a leap over something that does not rest but runs away. [Pushing off of water, itself.] On the swimming of animals of long shape such as eels and similar. On the manner of swimming against currents and great falls of the rivers.*

When writing notes to himself, Leonardo must have been doing his very best to keep up with his own racing mind, working to finish one thought before leaping, often on the same page, to his observations, ideas, and speculations about an entirely different subject. Later scholars who taught themselves to decipher his script still find it challenging to keep up with these leaps, not least because following Leonardo on any one subject requires endless research among many different, discontinuous pages across many different volumes. He himself wrote:

Here I shall not consider the proofs, which will be given later in the well-organized work, concerning myself now only with finding cases and inventions, gathering them as they occur to me. Afterwards, I shall create order by putting together those [cases and inventions] that treat the same subject; therefore, for now, do not wonder or laugh at me, Oh Reader, if I make great jumps from one subject to another here.

Taciturn and difficult?

The master has been depicted in modern times as a sage and sometimes even aloof. Contemporaries tell a different story. Leonardo was charming and persuasive in debates in a very agreeable way. He told jokes and riddles; he sang and played improvisations on his musical instruments to entertain others. His theatrical sets and devices were designed to delight and amaze audiences. His conversations with learned men were respectful and insightful. These are traits of an extrovert, not a grumpy man.

He seems to have worked hard at this persona. His humor was often written down in his notebooks, perhaps even practiced, for maximum effect among those who were more aristocratic and educated than Leonardo, who owned no property and had no legitimate birth. His inclusion was at the pleasure of others, whom he depended on for commissions, housing, and payment. He generally kept his grievances in check, only boiling over when he felt particularly badly treated by broken promises. Even after a bitter lawsuit against his half-brothers over property left specifically to Leonardo by their uncle, he maintained contact with them and their wives. At his death, he left that hard-won inheritance to the same brothers, as they had all agreed in a mediated settlement, and also left them his remaining cash savings.

Although Leonardo was considered strange by others for being a vegetarian, explicitly writing for himself that man need not thrive by the deaths of other animals, he did not proselytize. Similarly, he chose not to drink alcohol but did not forbid it in his household, and he bought both wine and meat for his staff and pupils. Vasari repeats a charming story about Leonardo's practice of buying birds in the marketplace. Instead of keeping them for song or food, he set his purchases free to fly back to nature, where they had been captured.

If he had an enemy, it was Michelangelo. That great artist was a true competitor for fame and commissions, not a friendly competitor. Michelangelo was dismissive of Leonardo, openly citing his reputation for never finishing projects. He considered the older master to be a has-been. Once, when Leonardo was on a panel to select the location for determining where to place Michelangelo's sublime sculpture *David*, Leonardo advocated for an out-of-the-way corner. His proposal, clearly born of pique, was wisely overridden by the other, more objective panelists. The unpleasant relationship between these two outstanding artists was the exception rather than the rule for Leonardo. He was normally cooperative with other artists, stepping aside from some commissions when the client requested it or collaborating with others on difficult projects for a painting or sculpture. An angler decides alone whether the most important goal that day is catching a lot of fish or one big one, exploring new water, or trying out a new technique that might catch nothing. Leonardo must have enjoyed his studies in the field and at his work table, where his own standards and opinions were the only ones that mattered.

Heretic?

Leonardo believed in God; of that there is no doubt. He found evidence of God everywhere in nature. Church doctrine appears to be a somewhat different matter. But consider his predicament: most of his paintings were commissioned by churches, and many of his greatest artistic innovations were achieved in depictions of religious subjects, from *The Baptism of Christ* to *The Last Supper* to *The Virgin of the Rocks.* In his sixties, Leonardo was patronized by the Medici pope's brother and given living quarters in the Belvedere in Rome. But he spent his days there on science, including hydraulic plans to drain a large marsh nearby, not on painting. At a minimum, professional and social courtesy would not have allowed any overt doctrinal challenges, even if he was so inclined. Nonetheless, he was a known homosexual when it was punishable by death under church law. He conducted illegal human dissections while in Rome, which eventually got him fired by the Vatican. In other words, he did not need to be outspoken against church teachings to be a heretic.

How can we square the circle of Leonardo's deep and lasting belief in God with his seeming disregard for the only Christian church then in existence in western Europe? The recipe for his creed seems to be a mixture of very ancient and very modern thinking, together with his own scientific skepticism. The Greeks and Romans believed that their gods had created a perfect world and universe. Long before Columbus, they knew that the world was round because a sphere was the mathematically perfect solid shape; why would the gods make our home any other way? Leonardo believed that God's work was also perfect in every respect, that every river, ocean, fish, animal, human eyeball, and muscle was designed exquisitely for its purpose. "Though human ingenuity may make various inventions," he wrote, "it will never devise an invention more beautiful, more simple, more direct than does Nature; because in her inventions nothing is lacking and nothing is superfluous."

Leonardo's grand objective was to understand how this wonderful world worked—all of it. He prized learning, knowledge, and virtue above all other things and believed all of it was within a man's grasp if he would work—tirelessly—at it: "Thou, O God, dost sell unto us all good things at the price of labor."

He paired this total faith in God's existence and perfection with an extraordinarily modern sensibility. It is common today for people to say that they are "spiritual but not religious." They believe in a higher power, even the one God of Christians, Jews, and Muslims, but do not "believe" in any organized religion. Sometimes they say that religion has co-opted God for its own purposes.

In Leonardo's day, the Catholic church owned vast estates, had its own army, exercised sovereignty well beyond the Vatican walls, and was sometimes ruled by popes corrupted by power, money, and even the flesh. Leonardo left Rome for the last time in 1516, just one year before Martin Luther nailed his ninety-five accusatory theses to the church door of Wittenberg Castle.

Leonardo did not rebel or openly express anything inflammatory against the Catholic church itself, beyond the riddles-as-prophecies he wrote about the foibles of most men, including clergymen. A couple of these are as follows: "Many will there be who will give up work and labor and poverty of life and of goods, and will go to live among wealth and splendid buildings, declaring that this is the way to make themselves acceptable to God. (Of friars and churches.)" And, "A vast multitude will sell, publicly and unhindered, things of the very highest price without leave from the Master of those things, which never were theirs nor within their power; and human justice will not prevent it. (Of the sale of Paradise.)"

However, Leonardo was critical and dismissive of all philosophers and of anyone who merely repeated the conventional wisdom of books or lived solely inside their heads. As repeatedly stated here, he valued observation and experience over a knowledge of traditional scholarship, frustratingly written in the Latin that he did not understand well. Perhaps it is only natural that he would believe in God, whose works he personally found over and over to be perfect, but not believe so much in whatever human teachings were not provable.

Vasari's assertion that Leonardo was something of a heretic (deleted in the second edition Vasari published eighteen years after the first) has some truth in it. While Leonardo certainly was not outspoken in his doubts about several important church teachings and doctrines, he recorded several thoughts and discoveries in his notebooks, which, if revealed, might well have seen him burned at the stake: "Here a doubt arises and that is: whether the Flood which came at the time of Noah was universal or not. And it would seem not, for the reasons which will now be given." And, "if you should say that it was the Deluge that carried these shells away from the sea for hundreds of miles, this cannot have happened since the Deluge was caused by rains; because rains naturally force the rivers towards the sea with the objects carried by them, and they do not draw up to the mountains the dead things on the seashores." And,

> The earth is not in the center of the Sun's orbit nor at the center of the universe, but in the center of its companion elements, and united with them. And any one standing on the moon, when it and the sun are both beneath us, would see this, our earth, and the element of water upon it just as we see the moon, and the earth would light it as it lights us.

Leonardo wrote: "The Sun does not move." When Galileo wrote the same thing in 1615, demoting earth's importance in the universe, it was condemned by the Inquisition as heresy. The second time he expressed it in 1632, he was tried and convicted, forced to recant, and spent the rest of his life under house arrest.

Finally, Leonardo's lack of belief in the immortality of the soul may have been the most heretical, but he must have kept it strictly to himself and his notebooks: "The soul's desire is to remain with its body, because without the organic instruments of that body, it can neither act nor feel."

Always focused on learning, he saw that a man's insights might give a certain form of immortality if they led to lasting knowledge or deeds: "Shun those studies in which the work that results dies with the worker."

No one followed this advice better than Leonardo himself.

THE RASCAL AND THE GENTLEMAN

STUDENTS, WORKMEN, AND ASSISTANTS PASSED IN AND OUT of Leonardo's household and workshops through the years. Some were talented, some were not; some worked hard, and some infuriated Leonardo by neglecting their work to go off bird hunting in the Roman Forum with members of the pope's Swiss Guard. But two people stayed and enriched Leonardo's life and ours.

The first was a ten-year-old boy, Giacomo Caprotti, who came into the household as a servant. His father is variously described as a poor tenant farmer or a man capable of paying his son's expenses to see his artistic talent developed. Little Giacomo was so mischievous that, in Leonardo's household, he quickly acquired the nickname "Salai," meaning little devil or trouble-maker. Leonardo immediately discovered the child to be a "thief, liar, obstinate, greedy." He stole money from a soldier's pants pocket while the soldier was trying on a costume Leonardo had designed for a court play. He sold a leather hide Leonardo intended for new boots and used the cash to buy candy. When Leonardo entrusted him with money to buy supplies, Salai spent it on himself. Leonardo was constantly bailing him out, admonishing him, and writing in notebooks about him with both frustration and amusement:

[When they were dinner guests] . . . this same Giacomo supped for two and did mischief for four, since he broke three table flasks, and knocked over the wine, and . . . on the 7th of September, he stole a silver point, worth 2.2 soldi, from Marco, who was living with me, and took it from his studio; and when Marco had looked for it for some time, he found it hidden in Giacomo's chest.

Why was Salai not sent home? He must have had a winning personality, more impish than vicious. Importantly, he was uniquely beautiful, with curly locks that made him resemble in person an idealized form of youthful beauty. Contemporaries who describe him at all consistently describe him as beautiful. Both in Leonardo's day and ours, there was speculation that a romantic aspect was important to their long relationship. Salai may well have been the model for many drawings and several paintings, including Leonardo's *St. John the Baptist*, in which the subject is notably androgynous. And though he was troublesome, Salai was loyal to his master for many years (as he certainly should have been). And he had those curls. Leonardo's portraits of women and images of men usually included magnificent hair in curls. So do his drawings of water currents.

Scholars say that Salai is not, however, the beautiful young man with full lips and curly hair who appears in so many of Leonardo's drawings and sketches. That figure appears in his work well before Salai makes his entry into the household, even before Salai was born. It is simply the artist's depiction of idealized youth. But the curls he had drawn and painted over and over again were a very apparent attribute of Salai. Leonardo had discovered that idealized youth in real life; the boy and the man must have been irresistible.

Salai arrived in 1490 and stayed with Leonardo for more than a quarter century. Leonardo dressed him in beautiful clothes and ribbons to enhance both his appearance and the dignity of the household. It may have enhanced Leonardo's own self-image and prestige to walk through Milan, Venice, Florence, and Rome in the company of such an attractive pupil.

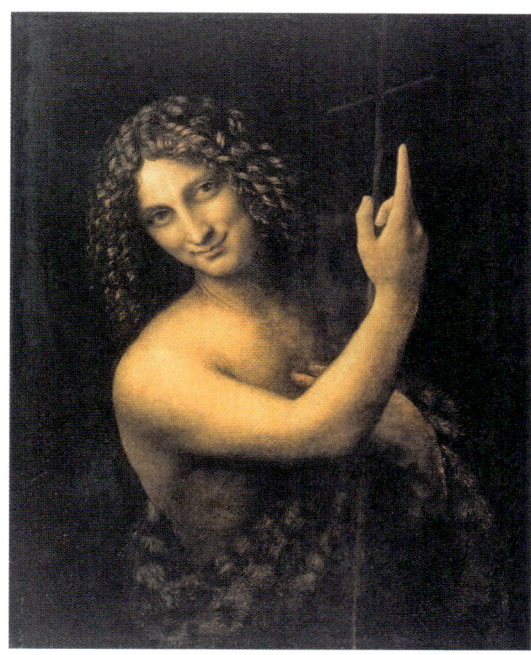

Leonardo, *St. John the Baptist*, ca. 1513–16, for which Salai was probably the model

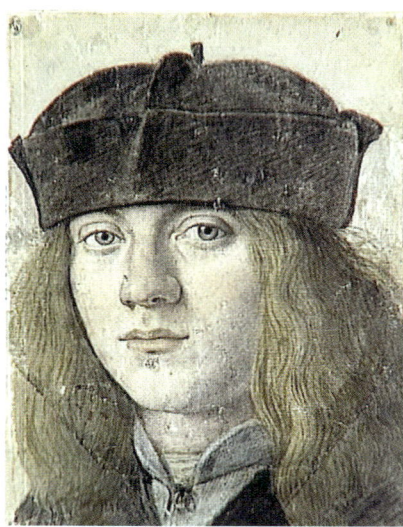

Francesco Melzi, possible self-portrait, ca. 1510

Francesco Melzi was born to be Salai's foil. The scion of a wealthy and noble family, he was fourteen years old when the by-then famous Leonardo was a houseguest of his father, Girolamo, in 1507. The Melzi family lived in a grand villa that still commands a fine view from the bluff over the Adda River and Martesana Canal in Vaprio d'Adda, nineteen miles east of Milan. Leonardo may have sought out the father because of the latter's work as a high-ranking military engineer on Milan's fortifications, which interested Leonardo greatly. Many of Leonardo's architectural designs that were actually built were for improving fortifications. The younger Melzi would become like an adopted son to Leonardo, the closest to family he had in the last third of his life, or maybe ever.

Young Melzi impressed Leonardo with his flawless manners and budding artistic talent. He took on the boy as a student, meaning Melzi was now a member of the family, the household, and the entourage. What we know of Leonardo today from his notebooks is largely due to this relationship, which led Melzi into a life dedicated to serving, then preserving, editing, and interpreting, Leonardo's mind, intent, and works. Melzi is the best supporting actor and the unsung hero of our story.

Leonardo immortalized scenes he observed from his corner room on the top floor of Villa Melzi in several drawings that beautifully show the complex river currents he observed, along with the river's impressive, craggy rock faces and scenes of daily life along its canal.

The currents fascinated Leonardo; to cross the Adda required a ferry boat that would not be swept away. Naturally, he redesigned the hull and rudder of the local ferries to work with the currents, using water as a sailboat uses the wind—a "reaction ferry"—to propel the boat sideways and make the crossing much easier for the boatman. The ferry at Imbersago, upriver from Vaprio d'Adda, still carries people (and up to one car) across the Adda today using Leonardo's design. Where Salai was poor, habitually dishonest, and never a

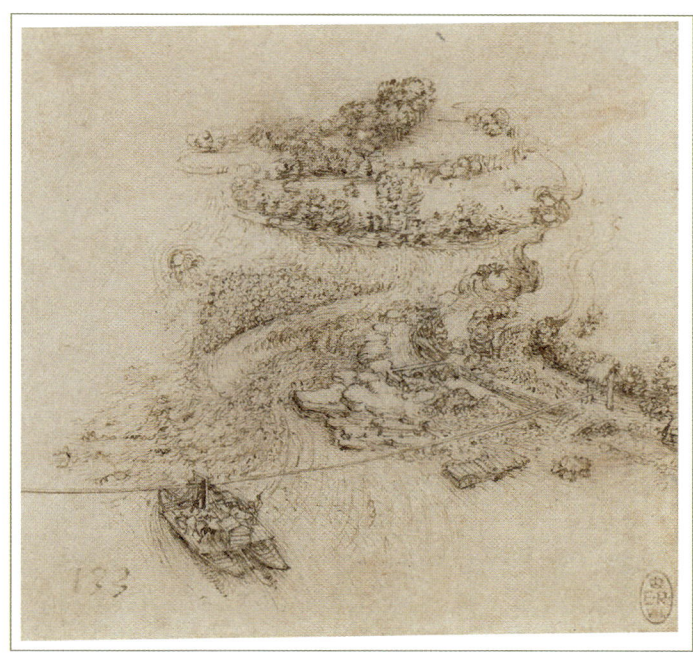

Leonardo, *A River Landscape* (Chain Ferry across the Adda River), 1513 (detail).
The currents fascinated Leonardo and required a ferry boat that would
not be swept away. The ferry upriver from Vaprio d'Adda still carries
people and one car across the river today.

great painter, Melzi was a dedicated pupil, upright, and never financially depen-
dent on Leonardo. What jealousies existed between the two boys and how
those might have played out is left by history to our imagination. But as they all
moved together around the cities of northern Italy, it was Melzi who became
Leonardo's assistant in increasingly important ways, particularly in helping to
organize the writings toward his planned books on water and on painting.

Years later—at the end of the master's life—Leonardo's deep attachment
to Salai resulted in an extraordinarily profitable transaction of some sort for

the younger man. They had probably fallen out a few years before, but Salai still made it profitable. Most accounts of Leonardo's last will and testament say that he left the *Mona Lisa* and the even more valuable painting *Leda and the Swan* (now lost) to Salai. But his will says no such thing. He left to Salai only a specific bequest of half of the vineyard near (and now inside) Milan that Leonardo had received from Il Moro, plus the house Salai had already built there. He left the other half of the vineyard to Vilanus, a relatively new servant. After twenty-five years together with Salai, Leonardo also refers to him as merely a servant in his will. Intriguingly, in addition to all of his books and notebooks, Leonardo left Melzi "*portraits appertaining to his* [Melzi's] *art and calling as a painter*" (emphasis added). Leonardo made specific bequests to others of money, furniture, and water rights, but not to Salai. Salai did receive a substantial amount of money from King Francis I in 1518, the year before Leonardo died, for several paintings, including the *Mona Lisa* and *Leda and the Swan*. Did Leonardo give the paintings to Salai? Did Salai represent Leonardo in making the sale, honestly keeping a large portion of the proceeds? If so, what happened to Leonardo's share? He did not bequeath to anyone as much cash as Salai had received for the sale of paintings in 1518.

When Salai was killed in a crossbow duel six years after Leonardo's death, he had the *Mona Lisa* and *Leda and the Swan* paintings in his possession, which were appraised and recorded. These were most likely copies that he had painted. Or had he somehow helped himself to the originals during the final year of Leonardo's life? Or, even richer, had he sold them to the king of France and then taken them with him when he left Leonardo? In other words, was Salai as an adult still that same thieving little devil Leonardo met more than twenty-five years before, now grown into an accomplished con artist?

WHEN DOES FISHING START?

WE WERE STANDING IN THAT GLORIOUS PRINT ROOM at Windsor Castle, discussing Leonardo's era and fishing, when Martin Clayton asked me, "When does fishing start?" I answered with some authority that there are no seasonal restrictions in most of the western United States but that I fish mainly after spring runoff, through the summer and fall, when the aquatic insects fish eat are most prolific. He replied that he actually meant when did people start fishing. And implicitly, I believe, could Leonardo have been an angler back then? Oh.

Certainly, prehistoric people caught fish with their hands and with spears and nets. Groups that lived near the sea fashioned barbed fishhooks from bone and wove plant fibers into hand lines. Archaeologists have discovered dip nets with long handles and cast nets with stone weights as old as ten thousand years in Sweden and the American Northwest. By 2300 BCE Egyptians were using rods with multiple lines and baited hooks; by four thousand years ago both Scandinavians and Egyptians were incorporating barbs on their hooks.

We do not know when fishing became recreation, but it clearly had by the time luxury in Rome met luxury in Egypt. Around 40 BCE, the final pharaoh

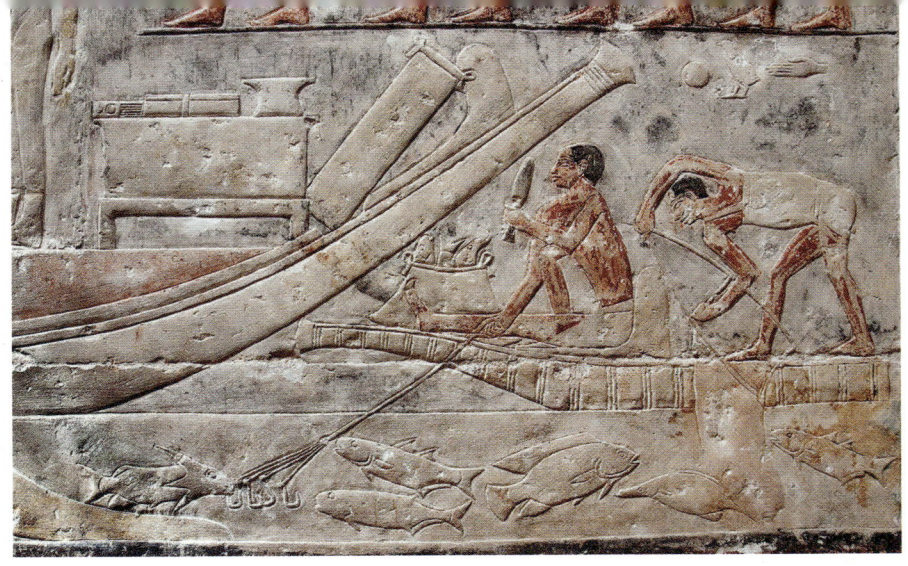

Stone relief of Egyptians fishing with lines and multiple hooks
from the large stone tomb of Vizier Kagemni at Saqqara, Egypt,
the cemetery city for the ancient capital of Memphis

was Cleopatra, of the line of Ptolemaic kings from Greek Macedonia. The
Roman writer Plutarch describes how she once again outwitted Marc Antony, who had gone to rule Egypt for Rome but was hopelessly seduced by the
wily pharaoh:

> One day Cleopatra took Antony fishing on the Nile on her fabulous barge,
> accompanied by a flotilla of smaller fishing boats. That day everyone on board
> pulled up a good number of Nile perch on their gilded hooks, everyone except
> Antony. Feeling humiliated in front of his lover and determined not to be
> skunked again, Antony devised a clever plan.
>
> The next day he secretly paid several of the fishermen in the smaller boats to
> dive underwater and place their own freshly caught fish on his hook. Plutarch
> tells us that over the next hour or so, Antony pulled up fish after fish. The

Egyptians on the barge marveled at the heap of silvery blue fish on the deck and wondered at the speed at which the perch were taking the Roman's bait.

Cleopatra immediately figured out Antony's ruse. But she feigned great admiration, exclaiming over what a natural fisherman he was. Declaring that his haul would surely be even more impressive tomorrow, she invited everyone back for another day of fishing.

The next day Cleopatra arranged her own cunning trick. As soon as Antony let down his line, some of her servants dove down out of sight. They attached a very large, very dead, salted fish from the Black Sea onto Antony's hook. Feeling the tugging on the line Antony quickly landed the heavy fish. As everyone stared at his catch it was instantly obvious that he'd hauled up a big fish that was not only dead and dried but not even from the waters of the Nile. As his biographer, Plutarch, comments, you can imagine the guffaws and hoots that ensued. Cleopatra, giggling mischievously, diverted Antony's irritation with flattery: "Better to leave fishing to us poor Egyptians—your game is conquering kingdoms."

This story is likely the first written reference to women fishing for sport, but hardly the last.

Within the Roman Empire, people were already fishing with artificial flies, mainly red yarn tied around a hook. The poet Martial, who wrote epigrams in Spain around 90 CE, offers this one: "Who has not seen the scarus rise, decoyed and killed by fraudful flies."

The first surviving written reference to fly-fishing for trout in a freshwater limestone stream describes fishermen in Macedonia, northeastern Greece. N. G. L. Hammond, a British secret agent and saboteur behind enemy lines in World War II and later a professor at Cambridge, quotes natural history writer Aelian around 200 CE: "They have planned a snare for the fish, and get the better of them by their fisherman's craft. . . . They fasten red wool . . .

round a hook, and fit on to the wool two feathers which grow under a cock's wattles, and which in color are like wax. Their rod is six feet long, and their line is the same length. Then they throw their snare, and the fish, attracted and maddened by the color, comes straight at it, thinking from the pretty sight to gain a dainty mouthful; when, however, it opens its jaws, it is caught by the hook, and enjoys a bitter repast, a captive."

Europeans were not the only ones to invent fly-fishing. Independently, Japanese peasants began fishing with light flies and long rods as early as the twelfth century. By the late 1500s the technique had been adopted by samurai warriors. They used it as a way to train their minds in patience, precision, and concentration, and their bodies in subtle arm movements and balance without violating bans on directly practicing martial arts.

Most well-read fly-fishers will name *The Compleat Angler*, by Isaak Walton, as the first book about our sport. Fly-fishing societies frequently have names like the Isaak Walton League in honor of the author of this disappointingly boring book, still in print today, long after its first edition was published in 1653.

But nearly two hundred years before Walton, an English gentlewoman named Dame Juliana Berners wrote *A Treatyse of Fysshynge wyth an Angle* after she had left courtly life for a Benedictine nunnery. Fishing for recreation was a fashionable group sport at court, and Dame Juliana does not seem to have left it behind. It is delightful to read this "pamphlet," published in 1496—or, I should say, to decipher it. It was published before uniformity of spelling in English was part of the culture and also before printers had stopped using an *f* shape to indicate an *s* sound. Consequently, the modern reader stumbles a bit but ultimately has little trouble understanding Dame Juliana's fishing tips based on her firsthand experience.

(Although modern English translations of ancient and foreign fishing references use the word "hook," this essential piece of fishing gear was called an "angle" by Berners. For millennia fishhooks were made of sharpened shells,

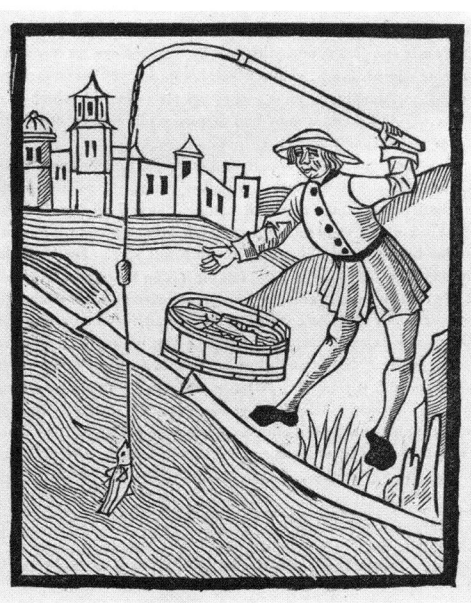

Illustration from Juliana Berners's
*A Treatyse of Fysshynge
wyth an Angle*

branches with sharp thorns, animal bones, antlers, ivory, copper, bronze, and iron. Most had a sharp barb forming an angle to the main curve of the hook. Thus people who fish are called anglers. We could have been called hookers, but I guess that term was already taken.)

Dame Juliana discusses which artificial flies are best to fool which species of fish in the rivers near the convent where she was the abbess. She calls flies "dubs" in the same sense we refer today to "dubbing," which is the animal fur, hair, or yarn wrapped with thread around a hook shank to form the body of an artificial fishing fly. She explains in detail how to make a good long "rodde" from tree branches and "lyne" from braided horsehair that would float atop the water. And she describes how to make flies for different species of fish, which a significant portion of industrious and dexterous anglers still do today with great satisfaction. Most amazingly, she describes in great detail how to make a "hoke"—even a variety of differently shaped hooks—for the flies/dubs she will catch fish with. If there are fly-fishers outside Japan making their own hooks or lines today, I would love to meet and fish with them.

Dame Juliana was a near contemporary of Leonardo da Vinci. Her life flourished around 1460 to 1490 and his from around 1482 to 1518. While he never visited England nor she Italy, I speculate that they would have gotten along. Like Leonardo, she was a keen observer and highly experienced. Her acerbic comments on the habits of men and her management of the nuns in her care would likely have made her a good conversation partner for Leonardo.

Both Leonardo and Dame Juliana looked down on men who were idle. Leonardo writes to himself:

> *I do not consider that men of coarse and boorish habits and of slender parts deserve so fine an instrument nor such a complicated mechanism as men of contemplation and high culture. They merely need a sack in which their food may be held and whence it may issue, since verily they cannot be considered otherwise than as vehicles for food, for they seem to me to have nothing in common with the human race save the shape and the voice; as far as the rest is concerned they are lower than the beasts.*

Dame Juliana was equally emphatic in this regard. She buried her pamphlet on fishing within her longer work on rich people's sports like hawking and hunting, so that the coarse classes would not be privy to her knowledge of effective fly-fishing. She orders her noble readers to fish wisely, not greedily; to follow rules of etiquette in their fishing; to not trespass or fish in the waters that belong to a poor man without his permission; to not open a gate without closing it behind themselves; to not steal fish that another man has caught and kept in a weir or pond; and to keep the techniques for catching fish (which she "teacheth and showeth" in her book) away from those who would violate these good standards, thereby ruining both the sport and the fishery. Like Leonardo, she was four hundred years ahead of her time, he as an engineer and she as a conservationist and ethical angler:

> *Here folowyth the order made to all those whiche shall haue the vnderstondynge of this forsayde treatyse & vse it for theyr pleasures.*

Ye that can angle & take fysshe to your plesures as this forsayd treatyse
techyth & shewyth you: I charge & requyre you in the name of alle noble men
that ye fysshe not in noo poore mannes seuerall water: as his ponde: stewe: or
other necessary thynges to kepe fysshe in wythout his lycence & good wyll.

Even in the "good old days" before crowds, Dame Juliana comments on the meditative aspects of fly-fishing, forgoing covetousness, fishing for the health of body and soul. Don't go with a lot of people, she tells us; go for your solace, presumably from daily life's cares:

Also ye shall not vse this for sayd crafty dysporte for no couetysenes to the
ncreasynge & sparynge of your money oonly but pryncypally for your solace &
to cause the helthe of your body and specyally of your soule. For whanne ye
purpoos to goo on your disportes in fysshyng ye woll not desyre gretly many
persones wyth you, whiche mighte lette you of your game.

To paraphrase the following passage: do not take too many fish at one time; that destroys the sport for yourself and others. Covet no more than a sufficient mess of them. Busy yourself in taking care of the fish, that they be nourished. To me, this implies stream improvement, environmental conservation, preserving adequate flows of cold water, and preventing pesticides from destroying the insects fish need.

Also ye shall not be to rauenous in takyng of your sayd game as to moche at
one tyme: whiche ye maye lyghtly doo yf ye doo in euery poynt as this present
treatyse shewyth you in euery poynt, whyche sholde lyghtly be occasyon
to dystroye your owne dysportes & other mennys also. As whan ye haue a
suffycyent mese ye sholde coueyte nomore as at that tyme. Also ye shall besye
yourselfe to nouryssh the game in all that ye maye.

As noted, she did not trust coarse and idle men to follow her orders. The final paragraph reads:

And for by cause that this present treatyse sholde not come to the hondys of eche ydle persone whyche wolde desire it yf it were enpryntyd allone by itself & put in a lytyll plaunflet therfore I haue compylyd it in a greter volume of dyuerse bokys concernynge to gentyll & noble men to the entent that the for sayd ydle persones whyche sholde haue but lytyll mesure in the sayd dysporte of fysshyng sholde not by this meane vtterly dystroye it.

In the world of fishing, facts are accepted as malleable. Most eighteen-inch fish are probably seen on a tape measure to be fifteen to sixteen inches long. If Descartes had been an angler, he would be remembered for saying, "I fish, therefore I lie." Dame Juliana's authorship of the *Treatyse* was accepted for centuries, but some now question whether her name was Berners, Barnes, or something else; whether she was truly the daughter or wife or widow of a nobleman; and even whether she existed at all. Her identity and gender, which are each fascinating to those of us imagining this nun fishing and writing when Leonardo was a boy, are neither provable nor disprovable.

Somebody wrote the *Treatyse*. No one has yet proposed an alternative author to Dame Juliana, based on clues of location, profession, or trade. I remain open to persuasion that I have been snookered by traditional myth, but for now I choose to believe.

As to that initial question of whether Leonardo could have been a fly-fisher or even an angler, I have finally settled on an answer: he could have been, but even I admit that there is not a single historical hint that he was. Rather, it is Leonardo's potential to guide us to fish in the currents, to teach us to observe in new ways, and to appreciate every element of fishing that I know is real.

THE BIG DIG

FLORENCE AND PISA JUST COULDN'T GET ALONG. Geography yoked them together, but history pushed them apart. Pisa had been an important coastal power and commercial port during Roman times, complete with its own navy. It was located on the northwest coast of the Italian peninsula, where the Arno River met the sea. Over the centuries the port silted up, gradually moving the coastline a few miles away to the west of the city. But Pisa survived as a prosperous republic and was the dominant city of Tuscany in all respects in the twelfth century, greater at that moment than regional rival Florence or naval rival Genoa.

After a disastrous defeat of its naval fleet by Genoa in 1284, Pisa went into decline. Meanwhile, Florence was rising in importance in the centuries before Leonardo was born. It was located about fifty miles inland, upstream on the very winding Arno. To reach the broader world, Florence's people, textiles, and other products mainly went through Pisa; transportation by water was much more efficient than by land, and Pisa stood astride the water route. Florence resented Pisa as much as Pisa resented Milan and Florence. The nearly constant tensions within Italian states and their rulers also made Pisa valuable to Florence's enemies. It came under successive domination by France, the Holy Roman Empire, and eventually the Duchy of Milan, which bought Pisa in 1399. For one ruler to buy a city and the right to tax its inhabitants from another ruler is as medieval a transaction as I can imagine. It was also a humiliating position for a formerly proud city-state and sometime republic.

Just seven years later, Pisa revolted against Milan. Florence used the chaos to conquer Pisa, first by brutal siege and starvation and ultimately by bribing the captain of the guards to open a gate in the city's formidable walls. So, Milan sold conquered Pisa to Florence; Pisa was now ruled by its bitter enemy of centuries. Florence proceeded to tax and punish Pisans in order to suffocate the city's economy. The enmity between the cities naturally worsened. This was the Tuscan political environment into which Leonardo was born and lived as a young Florentine.

Around 1490, while living in Milan, Leonardo began toying with the idea of making Florence an inland port by diverting the Arno from its natural and winding course into a new, straighter, man-made canal well to the north, through a tunnel in the mountains, and around pesky Pisa. Yes, it would require a vast amount of labor, technology, and money, but Leonardo was thinking big, and the goal would be transformational for Florence— economically worth all the costs if it could be achieved.

This was a fairly idle dream of the Leonardo da Vinci who had no real influence in Florence or anywhere else. But two events made the dream significantly more relevant. First, in 1494, France invaded Italy to claim the southern Kingdom of Naples for the French king. With French backing, Pisa refused to stay bought: it revolted against Florence.

Five years later the French army captured Milan, sending Il Moro fleeing and thereby depriving Leonardo of his patron and livelihood. The corrupt Borgia pope was allied with the French; his aggressive, illegitimate son Cesare entered Milan. Cesare offered Leonardo a position as his military engineer, but Leonardo declined and left the city. He wandered, working as a freelance artist in Mantua and as a hydraulic engineer on rivers and canals in Venice. In 1500 he drifted back to Florence after eighteen years away.

His home city was still trying to recapture Pisa from the French, five years on, with no success. That ongoing war was expensive, and Florence's economy was suffering. With increasing difficulty exporting wool and silk downriver to the sea and importing grains up the Arno, the possibility of famine was real. A canal around Pisa would prevent this from ever happening again.

To seriously complicate this already complicated mess, the pope had loosed Cesare on a ruthless and successful land grab up the peninsula. He conquered cities and territory for the sovereign Papal States, ruled by his father, and carved out territories for himself. Florence was in his sights when he stopped at Imola, a small fortress town just fifty miles away. Florence had neither an army nor sufficient mercenaries to defend itself.

Enter Niccolò Machiavelli. This ambitious poet, seventeen years younger than Leonardo, had risen to the position of second chancellor in the Florentine government and had proven adept as an adviser, diplomat, rapporteur, and administrator. Just thirty-three years old, Machiavelli was dispatched to Imola to bargain with Cesare to leave Florence alone. Cesare was both charming and ruthless, which may have been a factor in Leonardo's previous rejection of employment; we do not know. Nor do we know why Leonardo now reversed himself and went to work for Cesare as his general architect and engineer. In Imola and elsewhere, he was put in charge of assessing and surveying defenses, rebuilding fortresses, and improving walls, canals, and waterways.

Leonardo's survey *Map of Imola* shows his observations about how river currents are "reflected" back and forth across riverbeds, cutting deeper channels alternately on each bank. He writes in his notebooks about how to work with this phenomenon on weirs, stream alterations, and reclamation. The darkest shades of blue represent the fastest currents and deepest water. Just

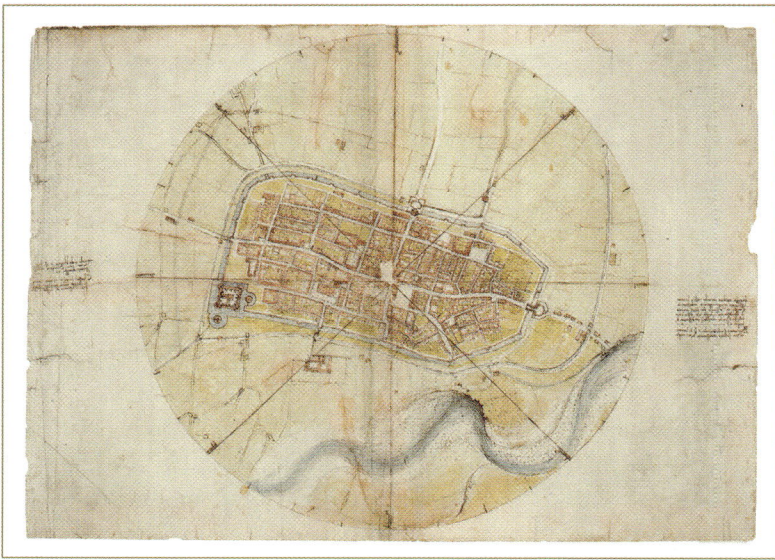

A Map of Imola, which Leonardo drew for Cesare Borgia in 1502,
pacing off each measurement on foot and demonstrating both his
engineering precision and his aesthetic sensibility

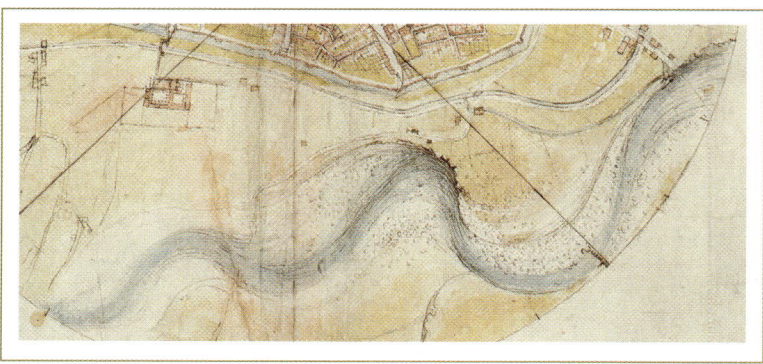

Detail of *A Map of Imola*. The depiction of the river's winding
course within its wider bed and the currents within the river
exhibits the knowledge of hydraulics Leonardo developed in Milan
and described extensively in his notebooks.

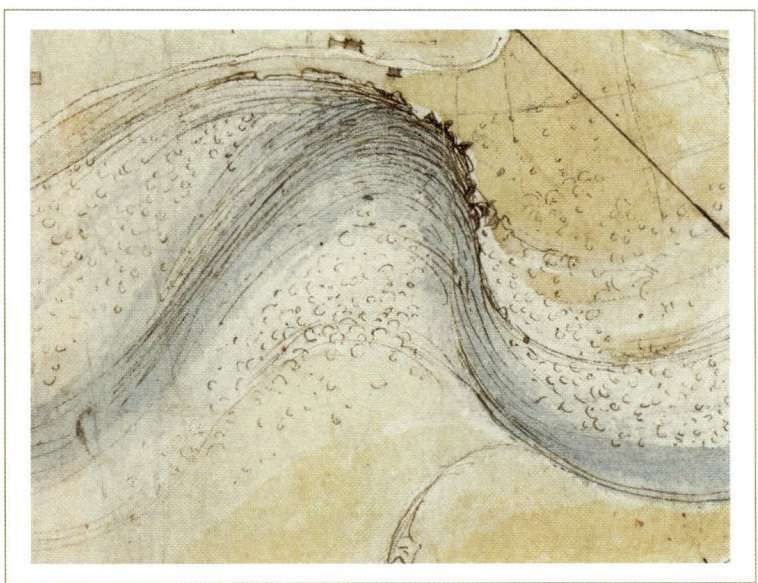

Detail of *A Map of Imola*. The best place to fish this section of the Santerno River would be at the top of the detail, in the slow water where the fast current swings away from the bank.

downstream, where the fast current swings away after bouncing off the bank, is often the best kind of place to fish. In that slow water, fish use less energy to hold and find food nearby. Leonardo's use of the now common bird's-eye view was extraordinary for his time.

So, Machiavelli and Leonardo were both in Imola in 1502 and 1503. These two very bright, curious, accomplished Florentines in this one small town, one working for Florence and one for its potential enemy, were bound to meet. Machiavelli reported everything to his bosses in Florence but never mentioned Leonardo. An intriguing theory for this odd omission, espoused by historian Roger Masters, is that Leonardo was recruited to spy for Florence from inside the aggressor's camp. We know from Machiavelli's reports that he had a well-placed informant in Cesare's inner circle, but the spy's identity is never revealed.

The second event that energized Leonardo's dream of rerouting the Arno occurred in 1504. Another native son of Florence, Amerigo Vespucci, electrified all of Europe when a private letter he wrote to the Medici from Spain was widely published. The letter claimed that some lands, which Columbus had discovered twelve years before, and which Vespucci had subsequently explored, were actually an entirely new continent. Vespucci wanted to see Florence join the race for the riches of the New World. But for the landlocked city to become any kind of seafaring power, it would obviously need control of its port—Pisa.

Leonardo and Machiavelli had each returned to Florence in 1503. Leonardo was well known there as a painter; few knew his talents as a military engineer or master of water. But Machiavelli knew. Leonardo had a plan to defeat Pisa. He proposed to divert the Arno's water into a swamp to the south of Pisa before the river reached the city. Machiavelli convinced the head of Florence's government to back the plan, which involved digging two canals, each deeper than the riverbed, and constructing a large weir to help divert the entire river south into the canals. This was not the young dreamer Leonardo in Milan but the master engineer. He calculated the vast amounts of earth that must be moved, the man-hours to move it, and the equipment he would need to design and build to assist the workers. A project engineer was appointed, and the digging began with two thousand laborers. The Pisans could only watch helplessly from the north bank.

But the engineer on-site made numerous, disastrous errors in executing the plan. He dug only one of the two needed canals. It was neither deep enough nor wide enough to drain the entire Arno out of its bed and into the marsh as Leonardo planned. At a key moment during the excavation, a huge storm sunk the boats of the Florentine soldiers guarding the essential diversion weir, drowning them and allowing the Pisans to move in and destroy the weir. The plan, as it was poorly implemented, was a total failure. The project engineer was blamed for both the failure and for the funds squandered when Florence could least afford them. Although Machiavelli and Leonardo did not shoulder the blame, both thought it wise to leave town for a while.

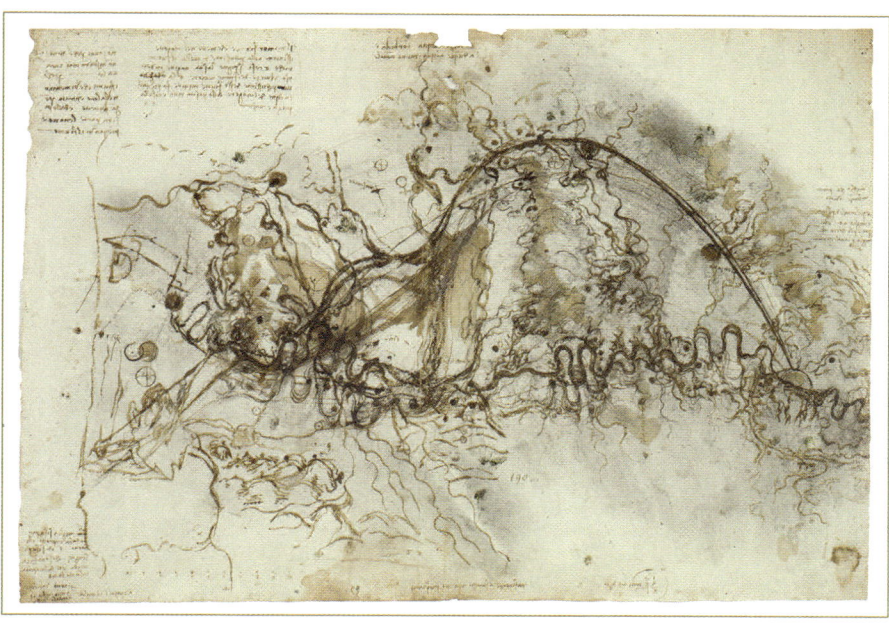

Leonardo, *Map of the Arno River and Tributaries*, 1503.
What appears to be a Jackson Pollock or modern abstract painting
is actually Leonardo's accurate map of the winding Arno River (horizontal across
the middle) and its many tributaries. Leonardo's planned peacetime canal to bypass
the river's bends and straighten the Arno forever is drawn as a smooth, dark curve
starting at Florence. Leonardo's grand route was never used for the Arno,
but four centuries later it holds the modern, divided Firenze–Mare autostrada,
carrying freight and people on tires just as he imagined it on boats.

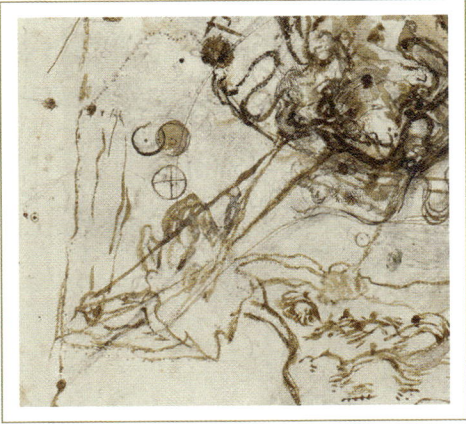

Detail of *Map of the Arno River and Tributaries*. The short straight line near the right edge in this detail represents Leonardo's proposed wartime diversion of the Arno River to bring rebellious Pisa to its knees. If properly executed, the deep double ditch would divert all the water from the Arno southwest into the swampy area shown somewhat heart-shaped, north of the hilly area.

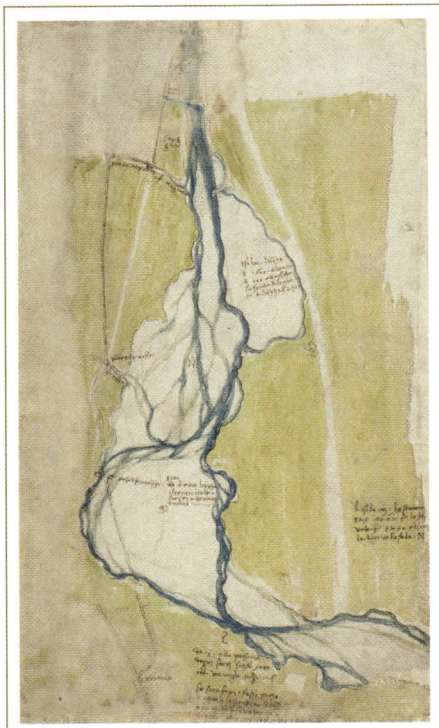

A Map of the Arno West of Florence, where Leonardo proposed to begin diverting the river around Pisa. Shown at low water with the riverbed channels exposed.

Not until 1509 did Florence finally reconquer breakaway Pisa. After fifteen years of contending with Pisa's stranglehold on exports and imports, perhaps Florence's leadership was ripe for Leonardo's biggest dream: the diversion of the Arno, making Florence an inland port for peaceful commerce with the world.

But it was too late. The rulers in Florence were soured by Leonardo's failure to complete commissions, and he had left for Milan three years before, in 1506. The more powerful French rulers in Milan were "asking" Florence for Leonardo's continued presence there, paying him a salary, reinstating ownership of property he'd received from Il Moro, and delaying his promised return. He never returned to live in Florence again except for a few months during the legal dispute with his half-brothers over their uncle's will. It is ironic that he is viewed today as a Florentine artist when the most productive two-thirds of his adult life were spent elsewhere, especially in Milan.

TROUT RODEO

AND THEN I FELL IN.

It started before that. We'd fished for hours, and the river was unusually low and clear. The Rio Grande near Taos is often high and muddy because the rain upstream can be unseen, far away, and not discourage you from hiking down into the Upper Taos Box. You don't even know the rain has happened or is happening, somewhere up in Colorado or over New Mexico's Red River, before it spills into the Rio, the fresh inflow full of mud and turning your fishing river to chocolate milk in front of your eyes. And ruining your day.

It wasn't like that. Too little rain everywhere made the river low, and the normally swift current was slower. The suspended dirt and small gravel had sunk to the bottom along the way, and you could see into the river much better than on normal days. Most important, you could walk out to some boulders that are unreachable on normal days; this day the water was low and the current was perfect.

I had caught a fair number of fish, but the big ones kept getting away, spitting the hook, or getting downstream and using the water's force against me, the current magnifying their size and weight when they turned a bit toward the bank and lost that sleek, fishy profile. It was late afternoon, and we still faced a climb out of the box, seven hundred miserable vertical feet up the switchback trail to get back to the flat canyon rim on top.

Boulder hopping on this part of the Rio Grande is tricky. We were high up in the river's course, so as Leonardo noted about the nature of rivers, the basalt down in that gorge is still in big chunks. Sometimes they're blocks with sharp edges, but usually they're smoothed out by water rubbing against those blocks for a few thousand years with fine grit suspended in it, like scouring powder. When you want to get closer to that

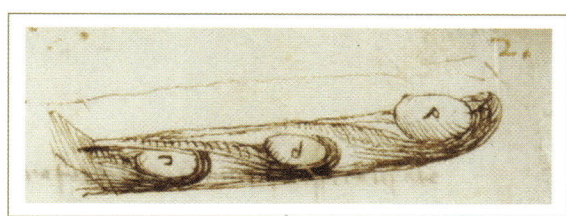

Detail of Leonardo's study of the effect of different obstacles on river currents

fishy-looking spot and the water is four or five feet deep, you have to leap a few feet across the water between the boulder you're on and the slick one you wish you were on. Nick does this all the time because the Rio is his favorite river, but I don't. That long hike back up the canyon keeps most sane people away, so the fish down there are bigger and more numerous and "easier" to fool with a fly.

I only go down there a couple of times a year, usually after the aches and fatigue of the previous trip have faded from my aging memory and I've forgotten my resolve to never go there again. Or because Nick says he was just down there and the fishing was great. Anyway, it was late in the day and I hadn't gotten a single large fish in the net. So, I followed him out three and then four large slick boulders away from the bank until we were near a big pool with good currents for any trout and plenty of depth for a big one. On the third cast, the drift was right, the indicator fly paused slightly, and I set the hook. Apparently, I was finally quick enough in doing so because a big, fat, beautiful

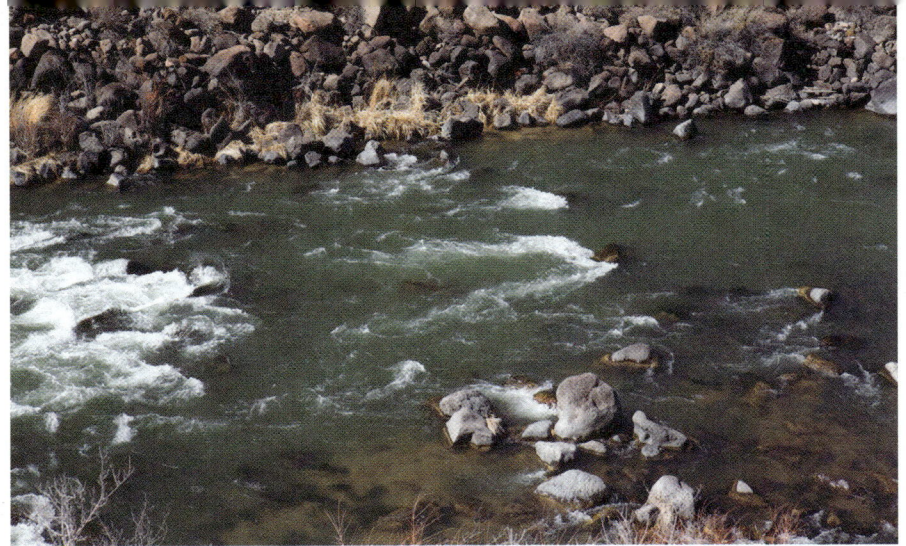

Typical currents

rainbow came shooting up out of the water; he and I were now joined by a thin line, if not quite at the hip.

Nick and I both hollered with excitement, and he begged me to not lose this one. (As if I'd been trying to lose all the previous big fish?) I did my best to keep the fish close, but my best wasn't good enough; the trout caught a fast current farther out and zipped downstream past me, stripping line from my reel as he ran. Nick yelled at me to stay with the fish, so I slid off the boulder into the water, only about three feet deep, and "ran" downstream to try to keep sideways pressure on the fish, my rod tip high and cigar firmly clenched in my teeth.

And then I fell in. I stumbled on one of the thousand rocks I couldn't see on the bed of that river. And a river is a river because it has a direction and a current. This is not a swimming pool or a pond but the Rio Grande, bound from high in the San Juan Mountains of Colorado to the Gulf of Mexico, twelve thousand feet lower in elevation. Carrying me with it. And it was cold.

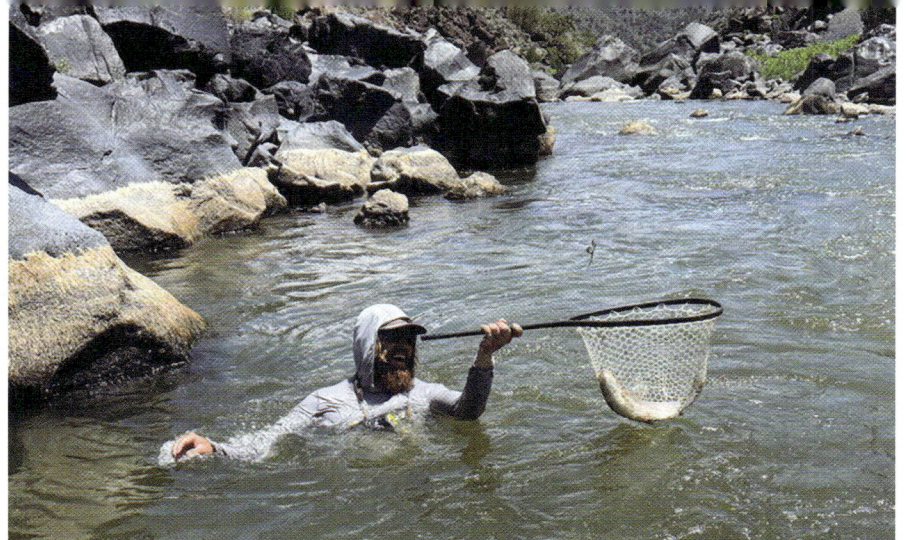

Success!

Each time I found a foothold to stumble to my feet, the current behind pushed me back down. Eventually I made it to vertical, got my rod tip back in the air, ignored the ruined cigar, and saw Nick. He was running with his net to try to catch up with the fish, which had neither gotten off nor taken more than about sixty feet of line. Now I was holding it from taking more, but the trout was sideways to the current and my arms were burning. Then Nick fell in. Completely soaked. He got up and tried to net the fish but missed. Then he tried again and my line went slack; the trout was in the net. Nick pushed against the current toward me, and I stumbled downstream toward him until we shared a boulder, admired the big rainbow, and released it for another day. Still laughing and dripping, Nick said, "Well, *that* was a trout rodeo!"

Leonardo thought and wrote a great deal about what we call geology today, particularly the formation of rivers, river bottoms, and mountains. Other than for mining, geology hardly existed as a science in his day. The earth was broadly thought to be only thousands of years old and created in its current form. The Great Flood was the one key physical event, and it was straightforward history, described clearly in Genesis. Leonardo thought past this: he observed the beds of seas that no longer existed; he analyzed how high mountain springs formed rivers, and how rivers carried away mountains in small pieces over extremely long periods of time. He thought and wrote about rivers turning high mountain boulders into rocks, turning lower rocks into gravel and then gravel into sand, all deposited on the beds of rivers where we wade.

He questioned how ocean fossils got to the tops of mountains. A single catastrophic flood would have receded to leave them in piles, like seashells on the ocean beach, but he found them exposed by rivers at different elevations. He was known for this peculiar interest; peasants brought him fossils to study when they found them, high in the mountains where they grazed their animals. He wondered how water got from the bottom of the sea to springs at high elevations based on Aristotle's theory that the earth, the macrocosm, had a circulation system like that of the human body, the microcosm. We still have vestiges of this way of thought: we call the Amazon rainforest "the lungs of the earth," absorbing massive amounts of carbon dioxide and producing massive amounts of sweet oxygen.

The Rio Grande Gorge in far northern New Mexico where Nick and I were fishing that day is ten thousand feet deep. I was standing in only three-foot-deep water, seven hundred feet down from the top, but the gorge started out ten thousand feet deep before it started filling back up with rubble. Once upon a time, long, long ago—30 million years ago—the earth tore open here. Most river valleys form between hills or mountains, but for miles on either side of this part of the Rio, the land is pancake flat, a high-altitude desert, hot and covered with chamisa and sage. Yes, you can see mountains reaching

Rio Grande Gorge in the rift valley

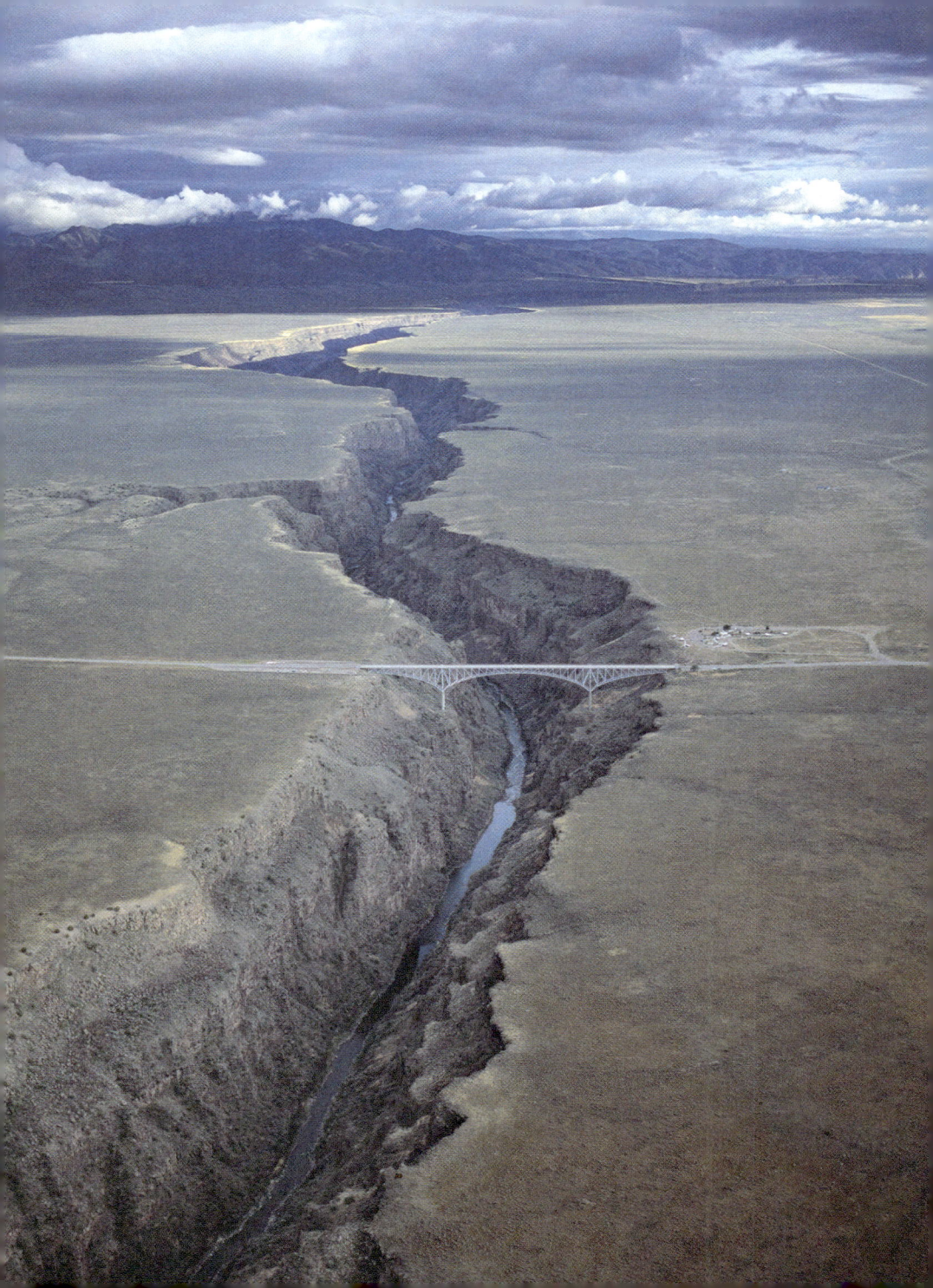

eleven thousand and twelve thousand feet above sea level both east and west as the river flows south, but the river didn't carve this canyon; the canyon caught the snowmelt and the boulders on their way down.

This is the grandest rift valley in the United States. It is a giant tear in the earth like the one the Jordan River flows through, the tear continuing under the Red Sea and onto the African mainland. If you drive from Santa Fe toward Taos, at one point you can look across twenty miles of flat desert scrubland and see the gash where the earth opened up, although you cannot see the river far below the open wound's edge.

Rain and snow in the high desert winter turn to ice and push open the little fissures between the rocks that form those canyon walls. Eventually a boulder is loosened enough to tumble down the wall into the steep gorge below and maybe into the Rio itself. When it does, the water's course in that one spot will change forever, or at least for a few thousand years. The current will be pushed around the rock, and the water just behind the rock will move more slowly. High water will dig out some depth around it and in the gravel just downstream. And a new place for a fish to hold will be born.

"WHY DID LEONARDO CUT [OFF] HIS EAR?"

TO SAY THAT THERE ARE A FEW MISUNDERSTANDINGS about Leonardo da Vinci is a great understatement. William Shakespeare is the only comparable giant of Western culture who left behind so many tantalizing clues and so few certainties. While there are many facts we do not and probably will not ever know, such as how many paintings he made, there is a lot we do know. And some of what we know is just not right.

The query quoted above is not from one uniquely confused soul; there are several that relate to Leonardo's ear. As far as we know, Leonardo's ears were just fine for hearing, and his mental health never deteriorated, even after a probable stroke a couple of years before his death at age sixty-seven. The confusion with the psychotic Vincent van Gogh, who lived three hundred years later in Holland, not Italy, and did cut off his ear two years before committing suicide at age thirty-seven, is hard to fathom.

Another incorrect "fact" relates to Leonardo's birthplace. Many tourists, including me, have visited Casa Natale, the house where Leonardo was supposedly born in Anchiano, an even smaller village two miles uphill from little Vinci proper. The house where Leonardo was not born was acquired, many

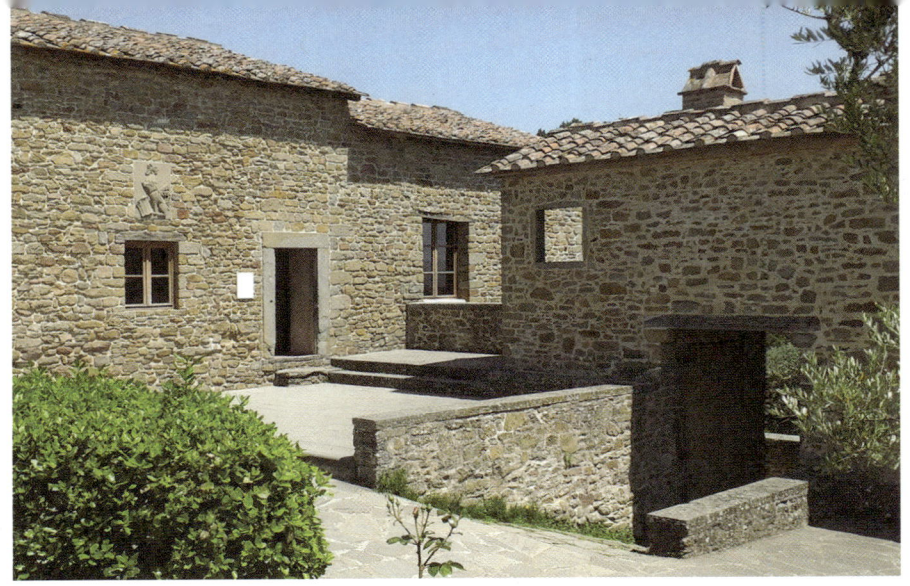

Casa Natale, which Martin Kemp calls "the house
where Leonardo wasn't born"

years later, by the da Vinci family, leading a romantic researcher to designate
this as his birthplace. Later research proved otherwise, but like most avid
anglers' tales, one must never let facts get in the way of a good story—particularly once paying tourists have made Leonardo's "birthplace" a major source
of revenue for the tiny village. In any event, the views across the rolling hills,
covered by olive groves, are worth the legend. Tiny streams below the steep
walls of Vinci and in the valley above Anchiano feed my imagination of where
little Leo might have first developed an interest in moving water. There was
machinery in the valley too; water mills powered olive-oil presses and might
have piqued his interest in mechanics. If not, at least the abundance of different kinds of butterflies would have entranced any observant child.

An important "fact" perpetuated since his own era is that Leonardo
wasted his time and therefore never achieved his potential as a painter. Those
who pronounced this verdict placed a high value on the art of painting,

especially compared with the engineering, science experiments, and rambling in the hills Leonardo did spend a lot of his time on. They also knew little or nothing of the written observations and scientific theories he disclosed mainly to his own unpublished notebooks. Had he spent more time painting, Leonardo probably would have produced more masterpieces, but I think our art-historical regrets would be better focused on the paintings he did produce that have been lost to the world in the intervening five hundred years.

Leonardo himself certainly did not share the opinion that he wasted his time. First, he had to make a living. He had no independent income from inherited properties, nor did he accumulate enough capital to purchase property. As he complained to the duke of Sforza in Milan when his promised pay from the duke's treasury was slow in coming (and it often was), he worked to live. His work as a hydraulic engineer and military engineer was more likely to be paid regularly than work as an artist, where project completion had to come first.

Celebrity artists in the fifteenth and sixteenth centuries were not paid the way star actors, athletes, and musicians are paid in the twenty-first century, when a couple of big hits or seasons can provide financial security for a lifetime. Indeed, there was no partial advance payment for the *Mona Lisa*, and Leonardo never personally received any funds for his sixteen years of work on the painting.

Second, although he wrote that "I know that many will say that this work is useless," Leonardo's scientific work was very important to him. Although he was both a product and a producer of the Renaissance, he foreshadowed the Enlightenment two centuries in the future, when confidence in the superiority of reasoning over faith would reach its zenith.

His science was written in his notes for the never completed "Treatise on Water" and was essential in producing the watery backgrounds and foregrounds of nearly all of his paintings and in the beautiful drawings of craggy valleys and the apocalyptic *Deluges*. These artworks simply would not have been possible without the time he "wasted" in the field and at his study table observing, calculating, and thinking deeply about the nature of water, skies, light, and nature.

It is often said that Leonardo was born ahead of his time. But his synthesis of many arts and sciences in fact speaks to a key difference between modern and Renaissance sensibilities regarding intellectual mastery. In our time, serious architects, historians, doctors, and hydraulic engineers follow rigorously prescribed courses of study before earning an advanced degree and a good job in their chosen field. They develop highly specialized vocabularies, skills, and thought processes unique to that profession. It is rare that an astrophysicist today will know a lot about physiology or painting techniques, because the demands of specialization today are just so great.

Most Leonardo experts say that he was, in fact, very much a man of his times. An educated person was expected to have a solid grounding in many fields, a liberal education. It is true that the bodies of knowledge in each field were much smaller than they are today, but Leonardo's contemporaries were at least conversant in several fields. What we call a Renaissance man today was a somewhat more common polymath then, a person with a wide-ranging knowledge of many subjects.

Leonardo knew more about more subjects than almost anyone in his own time because he worked so hard at it. He was excellent at making connections across fields we see as unconnected, because the pursuit of knowledge was what he lived for, and his curiosity and work ethic were unparalleled (if often derided). I have studiously avoided simply calling him a genius because it implies an ease of brilliance that belies how hard he worked all his life to gain what he knew.

Another oversimplification regarding Leonardo's artistic and scientific endeavors is that his science was in the service of painting or that his painting was done to prove his scientific conclusions. I believe attempting to resolve

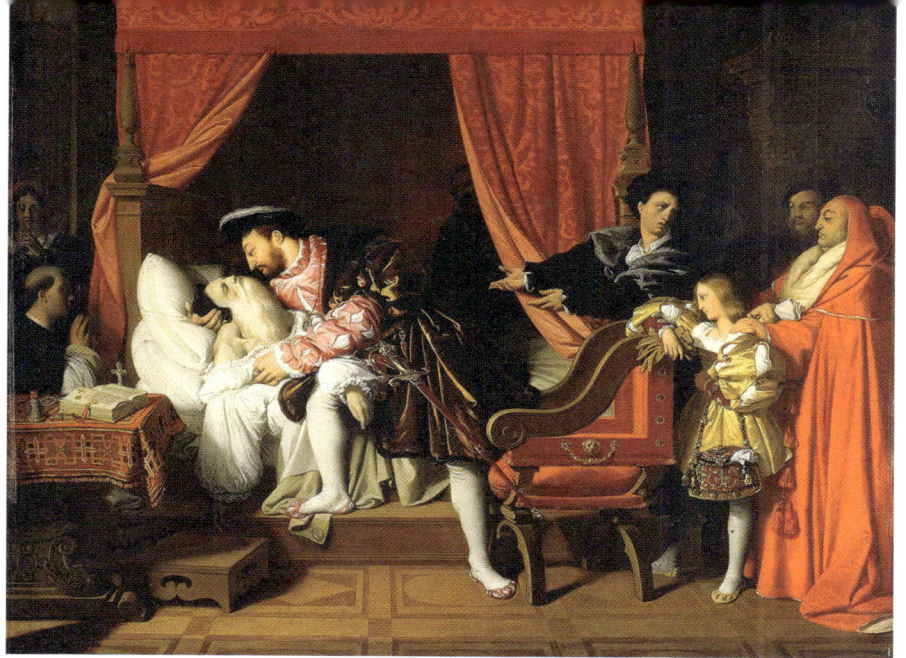

Jean-Auguste-Dominique Ingres,
The Death of Leonardo da Vinci, 1818

this question is as futile as determining which came first, the chicken or the egg. These two facets of his life enhanced each other in his output and his way of thinking; they both help define the unique figure he was.

A simple mistake that occurred eighteen months before Leonardo died is still making the rounds five hundred years later. An Italian cardinal and his secretary visited Leonardo at his home in France and reported that he had lost use of his right hand (presumably from a mild stroke). Not knowing that he was left-handed, the secretary reported that Leonardo could no longer paint although he could still draw. Some internet sites today state incorrectly that Leonardo wrote with his left hand but painted with his right and, incorrectly, that his disability (instead of his perfectionism) is why the *Mona Lisa* was left unfinished.

Several final mistakes in the folklore around Leonardo relate to unjusti-fied certainty about his final days and burial place, a fitting bookend to his mistaken birthplace. Vasari, who wrote a brief biography of Leonardo as one chapter in his *The Lives of the Most Excellent Painters, Sculptors, and Architects* (1550), describes a beautiful and dramatic scene:

> *At last, having become old, he lay ill for many months, and seeing himself near death, he set himself to study the holy Christian religion, and though he could not stand, desired to leave his bed with the help of his friends and servants to receive the Holy Sacrament. Then the king, who used often and lovingly to visit him, came in, and he, raising himself respectfully to sit up in bed, spoke of his sickness, and how he had offended God and man by not working at his art as he ought. Then there came a paroxysm, a forerunner of death, and the king raised him and lifted his head to help him and lessen the pain, whereupon his spirit, knowing it could have no greater honour, passed away in the king's arms in the seventy-fifth year of his age.*

Three hundred years later, this deathbed scene was dramatically por-trayed by the excellent French painter Jean-Auguste-Dominique Ingres. Unfortunately, this broadly accepted depiction of Leonardo's death in words and painting is likely false.

Vasari was an excellent painter, architect, and important biographer, but he was also a fine embellisher. It would not really do his story any good if the king of France was attending to administrative matters a two-day ride away, in Saint-Germain-en-Laye, where he issued an unrelated edict the day fol-lowing Leonardo's death (admittedly unsigned). Melzi, who was definitely at Leonardo's side at his death, makes no mention of the king's presence and knew Leonardo was only sixty-seven, not seventy-five. Nor is it likely that Leonardo expressed his "offense" against God and man for not painting more after a deliberate lifetime pursuit of science, nor that he experienced a death-bed conversion to extreme religiosity after a lifetime of quiet skepticism of dogma. It was what Vasari, a more religious and purely artistically focused man, would have wanted Leonardo to say; Vasari is scolding of Leonardo's rel-ative neglect of painting in his biographical chapter. But it is historical fiction masquer-ading as history, perpetuated and magni-fied by Ingres's dramatic and widely viewed painting in the Petit Palais in Paris.

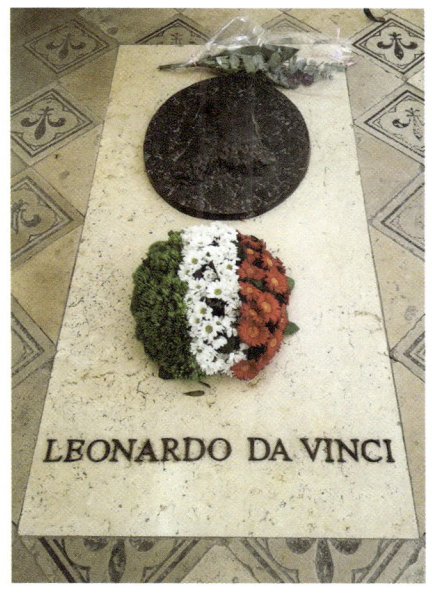

The popular but false accounting that Leonardo left the *Mona Lisa* and *Leda and the Swan* to Salai in his will was discussed ear-lier. So, I will recount here only the myth of the final resting place of Leonardo's remains. In strict accordance with his will, he was properly buried in the cloister of the Church of Saint-Florentin in Amboise, near where he lived his final years in France. That church was destroyed during the anti-religious violence of the French Revolu-tion. In 1863 a French poet and admirer of Leonardo excavated parts of the mass reburial ground and found fragments of his

Chapel of Saint-Hubert in Amboise, France, the final resting place of . . . someone

tombstone. He also found a skeleton with a large skull, which he romanti-cally decided must have belonged to Leonardo. These bones, whomever they belonged to, were eventually reinterred inside the Chapel of Saint-Hubert, also nearby Leonardo's chateau. Today, tourists visit his "grave" at Amboise, France, some seven hundred miles from his "birthplace."

A BREATH OF FRESH AIR

ROALD HOFFMAN IS AN IMP OF A MAN. No taller than I, he still retains the eastern European accent of his childhood. He smiles a lot and truly savors life. I met Hoffman just a few years ago at a wedding, where he danced every dance and all the women wanted to dance with him—which is particularly impressive since he was eighty-four at the time and already has a lovely wife. Like Leonardo, Hoffman overcame many hardships to attain greatness; he remains exceedingly modest. Even after winning a Nobel Prize, he chose to teach introductory chemistry to undergraduates at Cornell University. When he is invited to give lectures to experts around the world, he likes to pair them with talks for laypeople. He also writes opera and books of poetry and worked for years as a volunteer docent at Cornell's art museum to share his love of paintings and Japanese pottery with visitors. It would not be a mistake to call him a Renaissance man.

Nick Streit tells me that trout get stressed when the stream heats up in the afternoon because warm water holds less oxygen than cold water. Now that sounds like nonsense to me. Chemistry was my least favorite science class, but even I know that water is always H_2O—two atoms of hydrogen and one of oxygen, whether it is hot or cold. You can claim a thirteen-inch trout is six-teen inches long, but there are limits to stretching the truth, even for people who fish.

But Nick is right. Fish do not breathe the oxygen that is of the water, but rather the oxygen that is in the water—dissolved in the water, like salt or powdered lemonade dissolves in water. Water splashing over rocks or blown around by wind on its surface will grab and hold (dissolve) more oxygen out of the air than calmer waters will; calm water has less surface area and fewer air bubbles. So, when the stream water warms up on sunny afternoons, the trout will need to find oxygen in places where water interacts most vigorously with air, like riffles or plunge pools, even if that means greater exposure for them to birds and fly-fishers.

The reasons cold water holds more oxygen than warm water are submicroscopic. First, heat makes water molecules move around more, so there is more room for dissolved oxygen molecules to escape back into the air from warm water than when they're trapped among tightly packed cold-water molecules. Second, those tightly packed water molecules have more consistent interaction with the oxygen molecules, so the tiny electrical attraction can work more effectively.

Trout need the highest level of dissolved oxygen of all the river fish anglers target. They are the coldest of the cold-water fish. Pike can survive in waters with one-third as much oxygen as trout (and particularly trout eggs) need to survive. When we see bass or perch in our favorite rivers, those rivers may be getting too warm to support trout. Climate change might just take the trout out of your trout stream.

The amount of oxygen dissolved in water and available to fish is incredibly small. But it only takes one molecule of oxygen in every hundred thousand molecules of water to support a healthy fish population. Surely if a trout can find that needle in a haystack, it can find my fly!

English chalk streams and Pennsylvania limestone creeks, both famous for trout, also hold more oxygen because of the low acidity of those waters. But the low acidity is more important for healthy populations of the aquatic insects trout eat than it is to trout directly. Cool temperatures matter much more.

Leonardo's curiosity about water ranged from entire seas and rivers to a single drop. He may have been the first person to study and sketch the precise shape of a water drop and an air bubble; he wanted to understand how these most important of the four ancient elements actually work. A single drop of water is truly small. Scientifically, a drop is defined as one-twentieth of a milliliter (or one-twentieth of one-thousandth of a liter). When the eye doctor says to put one drop in each eye or the pharmacist says to put five drops into one glass of water, they need to be talking about the same amount of stuff; for safety, all the little bottles need to dispense the same size drop. But a molecule of water is a lot smaller; that one little drop contains an astounding 5,000,000,000,000,000,000,000 (500 quintillion) molecules of H_2O.

Here's the weird part: water grabs oxygen out of the air, and fish need it to do that. Leonardo could not have known about this; a molecule is way too small for even him to have observed, and the science we call chemistry didn't separate from alchemy until the 1700s. Joseph Priestly, the discoverer of oxygen and "father of chemistry," was a friend of polymath Ben Franklin, not polymath Leonardo da Vinci. (Coincidentally, when Priestly fled England for promoting religious toleration, he settled in central Pennsylvania just sixty miles north of those famous limestone trout streams.) Leonardo only knew the element of air; he did not know that oxygen was mixed up in that air with nearly four times as much nitrogen. Why and how does water grab the oxygen, which is so essential to the trout?

Hoffman explained that a water molecule is actually bent. I asked if he would draw a water molecule so you and I might understand what the chemists are talking about. Since his Nobel Prize is for using physics to determine how complex molecules are physically structured and shaped, my request was like asking Leonardo to sketch the *Mona Lisa* as a happy face. Hoffman being Hoffman, he agreed to do it.

In the two illustrations at right, you can see that within a single molecule of water, the small hydrogen atoms are not in a straight line with the larger oxygen atom. I have no idea why, but it turns out to be critically important to fish. Those two hydrogen atoms sticking out, perpendicular to the page, carry less of the dotted electron cloud than the water's one oxygen atom, making for a slightly positively charged place. (The water molecule as a whole has no net electrical charge.) For any water molecule that happens to be right at the water's surface, that tiny charged spot can act on an otherwise neutral oxygen molecule's electrons: the positive "ends" of the water molecule can pull on the oxygen molecules' electrons and "coax" oxygen into solution by forming hydrogen bonds. And there will always be an oxygen molecule right at the air's surface with the water (actually lots of them). Anglers might visualize this hydrogen part of the

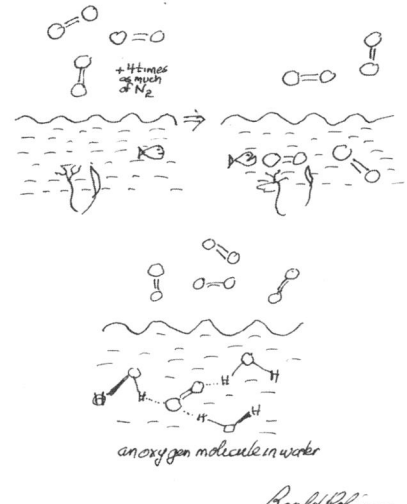

Oxygen molecules in the air
and underwater

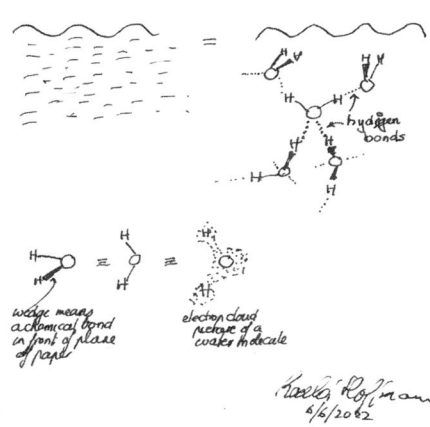

Oxygen molecules
dissolved in water

water molecule grabbing oxygen as a trout chasing an emerging insect right up through the stream's surface and grabbing it. Hoffman's drawing, then, shows an entire oxygen molecule under the surface, now tied to two water molecules (dissolved).

The last step is for the fish to pull the oxygen away from the water molecules and into its blood. Trout constantly take water in through their mouths and let it run out over their gills. Gills are made up of tiny filaments with large surface areas, filled with capillaries, like the 300 million alveoli humans have in each one of our lungs. On those gills, the oxygen-depleted blood passes next to the oxygen-rich water, and the fish exchanges carbon dioxide for oxygen, breaking the hydrogen bonds Hoffman shows as dotted lines.

Our lungs do the same thing with air, but usable oxygen is freely floating in the air, whereas it is bound to the water. If we fill our lungs with water, we can't break the oxygen out like fish do; instead, we just drown and die.

Since there are so, so many molecules of water in a drop, even one hundred-thousandth of the number is still 5,000,000,000,000 (5 trillion) dissolved oxygen molecules in a single drop of oxygen-rich water. The miracle of a fish's gill is that it can pull that dissolved oxygen right out of the water and into its bloodstream. That is why the gills of a fish are so red.

The point of this story is that we can observe the likely places for fish to hold not only by looking for obstacles forming seams between fast and slow currents. Anglers cannot see oxygen, but we can figure out where trout are most likely to find it, particularly on warm days. If there is plenty of food, oxygen, and cover nearby, we are probably in the right place to fish.

FISHING IN FOUR DIMENSIONS

LEONARDO WONDERED HOW RIVER CURRENTS WORKED under the surface as well as on top. Sometime between 1508 and 1510, he thought to observe water's actions at different depths by building a rectangular tank with a large glass window on one of the long sides. Today he could walk into a tropical fish store and purchase a hundred-gallon fish tank, but no such store or tank existed then. Undeterred, he designed it himself and likely had it built by a ceramicist.

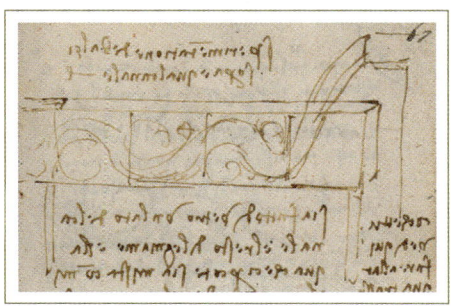 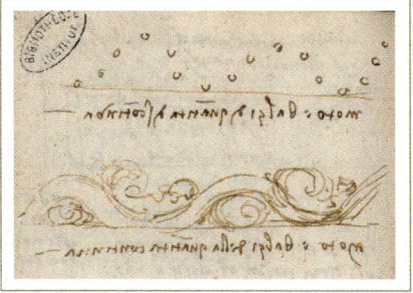

A glass-sided tank allowed Leonardo to simulate and observe currents operating underwater (detail)

The perfect tool for understanding nymph fishing (detail)

Using his tank, Leonardo studied the action of simulated river currents and ocean waves. He discovered how water and sediments could be driven to the bottom, only to bounce up and down again, moving light objects like insects around and shaping the riverbeds.

Fly-fishers learn pretty early on that trout feed mainly below the surface. As much as we love to place delicate imitations of adult mayflies with tiny wings on top of the water and see the heads of large trout break the surface to sip in the fly, we know that most of what they eat is the immature stage of aquatic insects—under the surface. These larvae or nymphs live far longer than the flying, mating stage: months or years underwater, compared with hours or days in the air. Nymphs may walk along the bottom, cling to a submerged rock or stick, or swim in the river, straining tiny amounts of nutrients out of the water as it goes by. Whether swimming with purpose or having accidentally been dislodged from a safe perch, the little drifting insect is liable to get swallowed by a trout.

As soon as an angler ties on a nymph or wet fly or streamer, she sacrifices the leading way to know that a fish took the bait: seeing the fly suddenly disappear from the surface in a ring of water. What she loses in surface-seeing by going down below, where most of the eating is taking place, she must compensate for with indicator flies, bobbers, sense of touch, and, most of all, visualization. Now she is fishing in her mind's eye. No one had a sharper mind's eye than Leonardo.

Let's assume that you have targeted a spot in the river where you think a fish would logically feed, based on all the ideas already discussed, and you have taken a guess how deep the bottom and hoped-for fish are. You have picked out a fly you think looks like what a fish would want to eat today. (There are entire books on that subject.) What happens after you cast a nymph toward that place is both simple and complicated. Simply, it either gets eaten by a fish or not. (Or it gets stuck on something on the bottom like a rock or a stick.) More complicated, it sinks at some rate of speed; gets buffeted around by a variety of currents and turbulence; gets dragged downstream by the line attached to your rod and the dry fly or a strike indicator on the surface (which is moving in a different surface current at a different speed); and it appears to a fish like either a delectable morsel or an alien invader. How can you improve the likelihood that the fly gets eaten?

Beyond hoping the fish is in a dining mood, the two keys to bettering your chances are for the fly to drift close enough for the fish to see it and bother to move a few inches to eat it, and to present it looking like a real, natural insect. These require you to imagine, plan, and execute your offering in all four dimensions: the right distance across the river that is in line with the fish so he doesn't miss seeing it entirely; the right distance upstream from the fish so the fly doesn't bonk him on the head, land behind him, or fail to sink far enough underwater for him to see it; and to dial in the complicating dimension of time, which constantly changes where the flies are (across, up, and down) and how fast they are drifting in the current.

Try to visualize this as if you are standing with Leonardo watching it happen in his glass-sided tank or as if you are scuba diving near a large trout. The fish is "sitting" two feet deep in three feet of water. The fly, tied onto the leader and tippet three feet behind a floating indicator or bobber, lands on the water and starts to sink. If the fly lands fairly close to the fish, or if the fly is fairly light, or if the current is fairly swift, or some combination of these, it will not sink deep enough in time for the fish to see it and eat it; it will drift unseen over his back.

If the fly lands too far upstream of the fish, or if the fly is fairly heavy, or if the current is fairly slow, it will sink so deep that the fish will not see it or eat it; it will drift by below him or get snagged on the bottom, creating other problems. Obviously, the "Goldilocks" solution we want is for the fly to land quietly, just the right distance upstream of the fish, in just the same current

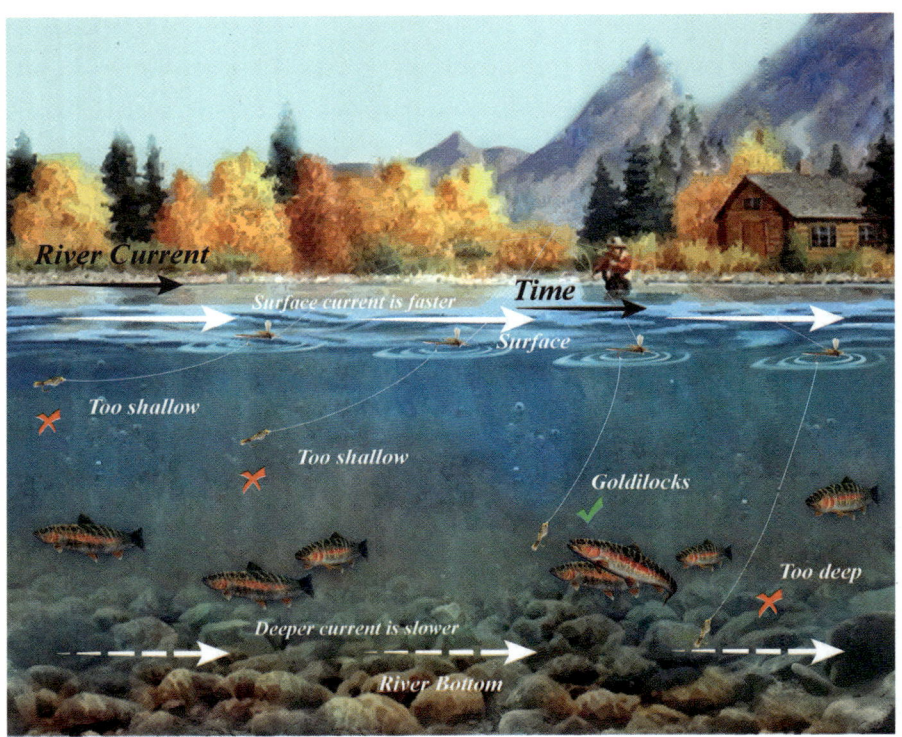

Fishing in four dimensions

line with the fish, and to be just the right weight. Combined with the speed of the current the fish chose to sit in, in the few seconds it drifts downstream, and it sinks close enough to his two-feet-deep face that he sees it and eats it. That's a lot of things to get right at one time. If you think in advance about all those things happening out of view, like Leonardo did studying them in his glass-sided tank in four dimensions, you will catch more fish.

If you start to take a bite of your favorite chocolate cake and it begins to move on the plate, I predict that you will instantly lose your appetite. Before you could even think about it, your animal instinct would put you off: something isn't right here! So it is with a fish and a fly that moves unnaturally in the water. Fish see so many insects and pieces of leaves and sticks moving past them all the time at the exact same speed as the water current carrying them that anything moving faster or slower will stick out like a sore thumb or a moving piece of cake. If you watch a marching band, you won't notice any individual member unless she is running ahead or falling behind. Even a fish that is fooled by a bunch of chicken feathers wrapped around a steel hook knows a fraudulent bug when it is out of step.

What causes our fly to zoom past a nice trout and cause an instant case of lockjaw? Drag. Drag is easy to see on the surface when a dry fly moves faster than the bubbles around it and creates a little wake in the water behind it. Usually that happens because a faster current between you and the fly is pushing and bending the fly line downstream, again playing the game of crack the whip, where the tip of the whip—the fly—is suddenly moving much faster than the slower current it is floating on. Visualizing this in advance by observing the currents closely, like Leonardo did, lets you plan how you will control the line, mending it upstream or downstream as needed to keep the surface fly moving at the same speed as the surface current.

Most guides and experienced anglers will feel the weight of the nymph, bouncing it in the palm of their hand before deciding to use it. I like to also drop it into the water immediately in front of me once I've tied it onto the tippet, watching how quickly it sinks. If I'm using two nymphs at the same time, that gives me an even better idea of how fast each fly will sink after it hits the water. That is what I'm visualizing as soon as I cast and they disappear into the uncertain depths.

Leonardo discovered that there are rivers within rivers. Almost always, the current below the surface is moving slower than on the surface. Near the bottom and along the edges of the stream, friction with rocks or banks slows the current even more, so the faster-moving top fly or indicator starts to stretch the tippet connecting to the nymph below. Over several seconds, the fly on the surface can begin to drag the nymph along faster than the unseen current below is flowing. A drag-free drift of a fly at the surface may actually be dragging the nymph down below, turning your delectable insect into that whizzing, repulsive alien. One solution for that problem is a longer tippet that takes more time to stretch tight. Unfortunately, the slack you have now intentionally created underwater means a slight delay in telegraphing to the top fly that a fish just ate the nymph below. The only solution I know to that problem is to watch the indicator fly very closely for the slightest unnatural movement and set the hook quickly to make up for the fraction of a second lost to the slack. When I foul hook a fish in the tail or catch him in the chin instead of the lip, I know I was too slow and he had already spit out the fly before I set the hook. All this gets better if I'm imagining the fly sinking and I've guessed correctly how deep the fish is in the water.

If the fish grabs the nymph and turns or runs, this is all easy because you will see the indicator move fast and hard. But big fish get to be big fish by being skeptical and taking flies gently, spitting out fakes quickly. We miss a lot of takes, probably by the biggest fish, when they take the fly subtly. Different guides have different guesstimates of what percentage of takes we never even notice, but it's a lot. It is part of what makes nymph-fishing challenging. If you correctly visualize when the nymph gets down to the fish, you'll be ready to set the hook quickly, and it gets to be much more fun.

To me, the speed at which a trout can spit a fly out of its mouth is astounding. Its lips and tongue are not nearly as flexible as ours, but it seems to use its mouth full of water to expel the fly in an instant. Other times a trout will close its jaw around the fly so slowly that a fast hook set pulls the fly right out of the water without even grazing any part of its mouth. When a fish jumps out of the water, the tension holding the hook in its mouth must vary so many times in that fraction of a second that the wily critter often manages to exploit some part of the change to dislodge it and spit it out. Finally, I am amazed when I have a fish in the net to see how it uses tiny mouth muscle movements to inch the barbless hook up toward its lips toward freedom. Trout don't write hit songs, but they have some very useful talents.

OH, WHAT A LOVELY MESS!

This will be a collection without order, made up of numerous sheets that I have copied up, hoping later to put them in order, in their proper places, according to the subjects which they treat.
— Codex Arundel, March 22, 1508

WIND WILL TANGLE YOUR LEADER. Panicked fish will wrap a dropper around the main tippet, creating knots too small and too many to untie. When your line hits a tree branch or a bush, a fly will wrap around it so many times and in ways that it is hard to believe it wasn't intentional. Years ago Claudia stopped going with me when I fished because she said all I did was change flies and untangle knots. Most of us eventually set a mental rule: if it's going to take more than X minutes to untangle, cut the mess off and start over.

Leonardo sat down in Florence in 1508 to sort out his notebooks—more than thirty years of thoughts, hypotheses, experiments, discoveries, lists, and more, "according to the subjects which they treat." He was only visiting Florence because of an inheritance lawsuit with his half-siblings; he had moved back to Milan three years previously, at the request of the French governor there. No one could foresee that Leonardo would spend the rest of his life increasingly patronized by the French crown. These would be productive and comfortable years.

He remained obsessed with water. Aristotle had named it as one of the four elements of the world, but what was it exactly? What rules did it follow, how could its behavior be predicted? How did it create and destroy life and land, rivers and oceans? How could an understanding of it be harnessed for mankind's benefit and protection? These and many other topics had absorbed a great deal of Leonardo's thinking; his writings and sketches were spread over many of the notebooks. It was time to turn them into a planned "Treatise on Water," along with separate treatises on painting, optics, anatomy, the flight of birds, and other important subjects about which he knew so much more than others.

He got down to work. The outline starts with instructions to himself:

Write first of all water, in each of its motions, then describe all its bottoms and their various materials, referring always to the propositions [conclusions] concerning the said waters; and let the order be good, as otherwise the work will be confused. Describe all the shapes that water assumes from its greatest to its smallest wave, and their causes.

> *Book 1 of the nature of water*
> *Book 2 of the sea*
> *Book 3 of subterranean rivers*
> *Book 4 of rivers*
> *Book 5 of the nature of the depths*
> *Book 6 of the obstacles*
> *Book 7 of gravels*
> *Book 8 of the surface of water*
> *Book 9 of the things that move on it*
> *Book 10 of the repairing of rivers*

Book 11 of conduits
Book 12 of canals
Book 13 of machines turned by water
Book 14 of raising water
Book 15 of the things which are consumed by water

As Leonardo transcribed previous writings and observations from his notebooks, it did get confusing. He tries to explain everything about water and how it interacts with air, land, and itself. He explains in such detail that one can get lost. When he lists the different ways water can move, he is more exhaustive than many Eskimo terms for snow, naming sixty-four terms in all! Martin Kemp rolls them out in translation: "Rebound, circulation, revolution, rotating, turning, repercussing, submerging, surging, declination, elevation, depression, consummation, percussion, destruction." Similarly, Leonardo lists 758 conclusions about what happens when different currents intersect, with a great variety of relative strengths and angles. The descriptions become mind-numbing. Words are not his best tool for expressing himself; drawings are.

As he explains one thing, he wants to explain more things in still more detail. The manuscript becomes more and more like an encyclopedic list of conclusions and less like a textbook a future aspiring hydrologist would want to read.

This is not to diminish the enormity of Leonardo's insights. For example, here he instructs himself to explain erosion, a continual process that was novel in concept—different from the general belief that God had created the earth just as we find it in the present:

How in the end the mountains will be leveled by the waters, seeing that they
wash away the earth which covers them and uncover their rocks, which begin

to crumble and, subdued alike by heat and frost, are continually changed into
soil. The waters wear away their bases and the mountains fall bit by bit in
ruin into the rivers . . . and by reason of this ruin the waters rise in a swirling
flood and form great seas. . . .

How loose stones at the base of wide, steep-sided valleys when they have
been struck by the waves become rounded bodies and many things do likewise
when pushed or sucked into the sea by waves.

There are sweeping, novel statements of geology, which hardly existed as a science: "Mountains are made by the currents of rivers. . . . Mountains are destroyed by the currents of rivers. . . . The summits of mountains for a long time rise constantly." The last of these is an especially radical idea; neither observation nor measurement at that time could possibly back up this assertion of mountains rising. How did Leonardo even conceive such a thing? Plate tectonics was first proposed in the twentieth century and only gained broad acceptance in the 1960s.

Leonardo's mind moved quickly across his fields of interest. Although he was trying to sort his ideas into "the subjects they treat," his search for a sort of Grand Unified Theory of Nature also allowed him to make insightful connections. Here is a very watery passage from what became the Codex on the Flight of Birds:

Why is the fish in the water swifter than the bird in the air when it ought
to be the contrary since the water is heavier and thicker than the air and the
fish is heavier and has smaller wings than the bird? For this reason the fish
is not moved from its place by the swift currents of the water as is the bird by
the fury of the winds amid the air; also we may see the fish speeding upwards
on the very course down which the water has fallen abruptly with very rapid
movements like lightning amid incessant clouds, which seems a marvelous

thing. And this is caused by the immense speed with which it moves which
so exceeds the movement of the water as to cause it to seem motionless in
comparison with the movement of the fish.

Of all the subjects Leonardo studied, he makes the most progress in sort-
ing out his writings on water and on painting. Around 1510 he focuses on
water in what would become the seventy-two-page Codex Leicester and
parts of the Codex Arundel. In them, he explains why we (anglers and others) find boulders and pocket water high up in mountain streams and more
smoothly flowing rivers with gravelly beds lower down. He explains scores
of ways in which various currents interact at the surface, deeper in the water
column, against the edges, and against the bottom. Particularly interesting to
fly-fishers are his illustrations of what happens to currents when they percuss
a variety of "obstacles" in pairs and groups, creating pocket water.

As he worked on his treatises, Leonardo saw the problems with his own
method:

I fear that before I have completed this, I shall have repeated the same thing
several times, for which do not blame me, reader, because the subjects are
many, and the memory cannot retain them and say, This I will not write
because I have already written it. And to avoid this it would be necessary, with
every passage I wanted to copy, to read through everything I had already done
so as not to replicate it.

By 1512, Leonardo had not fulfilled his dreams and plans to write comprehensive treatises on the many subjects he had spent much of his life mas-
tering. He was sixty years old and must have known by then that he was
unlikely to live long enough to complete all these projects. Each topic was

spread over many notebooks, and each notebook contained writing and sketches on many topics. More than twenty-five thousand pages were bound in "an infinity of volumes," according to one visitor in 1517 in France, without any consistent organizing system (and no Post-it notes). It amounted to a trove of brilliance and a lifetime of discoveries, all in a totally unusable form.

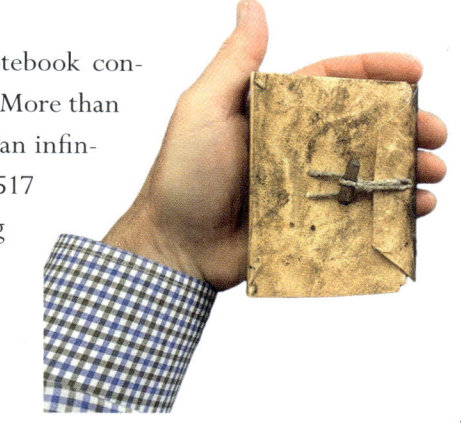

Facsimile of one of the notebooks Leonardo carried on his belt

Leonardo began to organize his many notebooks in a way that would allow this work to be completed after he had passed from the scene. Francesco Melzi, his able and loyal assistant since 1507, worked tirelessly with him to complete the work. Further complicating this task, Leonardo did not stop working on new ideas: his drawing, science (including water), and architecture continued. In fact, these were very productive years. Together, master and assistant spent twelve years trying to organize it all, but they failed. Leonardo left all of his writings and drawings to Melzi in his will, believing the younger man would faithfully work to complete the project. This was one of Leonardo's wisest judgments; Melzi spent the next fifty years working to organize and publish Leonardo's notebooks and papers, until he too died, in 1570. Melzi also failed. In the intervening five centuries, much additional work was spent on this project, and in some senses it has failed too. Why?

To this day, we celebrate the enormous scope of Leonardo's interests and expertise. Just the portion of his recorded thoughts in written and drawn forms that have survived, perhaps one-fifth of the original total, far exceed the archive of any other artist of the Renaissance. Yet they are a mess. It is

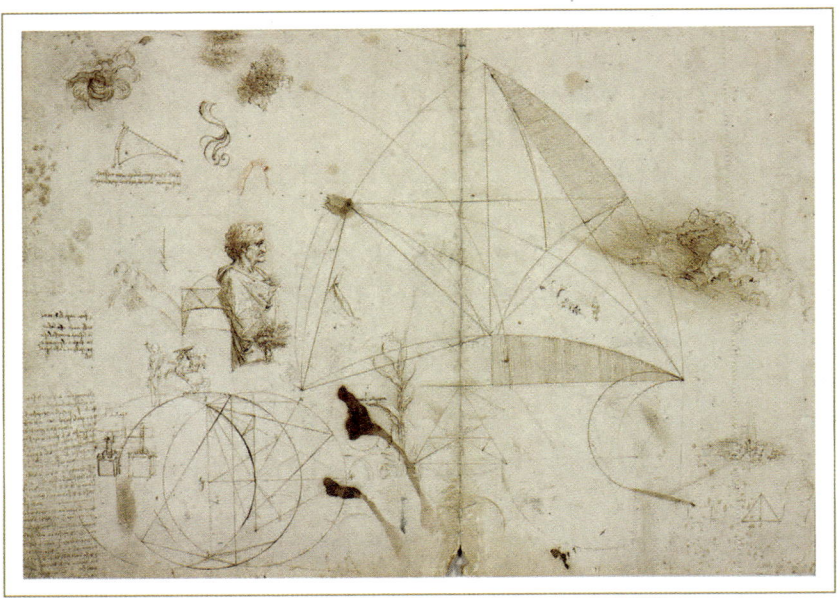

Leonardo, *A Sheet of Miscellaneous Studies*, ca. 1490

exactly this enormous scope of interests that is the problem; he thought and wrote and drew about them all more or less at the same time and often on the same page.

Martin Clayton, head of prints and drawings at the Royal Collection Trust and an expert on Leonardo drawings, wrote about the sheet above in his book *Leonardo da Vinci: A Life in Drawing*:

> This large sheet has been seen as typifying Leonardo's effortless movement
> from subject to subject in his drawings, but few sheets cover such a range of
> material. First to be drawn were the two large geometrical diagrams, in which
> Leonardo played around with circles, arcs and triangles—the transformation
> and equivalence of areas bounded by such lines was an abiding hobby. Around
> these diagrams he fitted small studies of many of the subjects that were

bubbling away in his mind in the years around 1490: a rearing horse with rider echoing the Sforza monument, and a standing warrior; two sketches of a screw press; an old man in profile, his cloak merging with a study of trees; and a smaller geometrical diagram that morphs into mountain peaks. Below centre is a meticulously drawn blade of grass, and at top left an arum lily and two sketches of foliage. At centre right is a bank of strongly lit cumulus clouds, as studied in drawings and notes to the end of Leonardo's career. Below is a sketch of water falling into a pool, though the scale is hard to gauge—it could be a puddle, or a vast cloudburst comparable to the late Deluges.

Melzi and his followers confronted thousands of pages and thousands of drawings. Should they be organized by topic, as Leonardo and Melzi worked to do? That could mean unbinding the manuscripts and even cutting up sheets and making separate piles of fragments that related to geometry and anatomy and optics and water and machinery.

That approach would mean quickly losing track of the order in which each drawing and page of notes was produced, thereby losing the evolution of Leonardo's thinking and learning. Or should we preserve the chronological order to understand that evolution in his thinking at the expense of consistency of topic?

To make the puzzle considerably more difficult, after Melzi's death, his son Orazio, who inherited the family villa with a room full of these volumes, did not care about or value them as an archive or much of anything. Thirteen books were stolen by a family tutor, and when an honest monk sought to return them, Orazio said he had plenty, so just keep them. Souvenir hunters were allowed to take drawings or whole volumes without so much as a record made of what was disappearing. Progress made by Leonardo and then Melzi, over a combined sixty years, evaporated. Collectors like Vasari began to purchase notebooks or leaves of notebooks for their drawings, many of which they individually reorganized in ways they found interesting or logical, and rebound them into new manuscript books (known as the codices).

The most important of these collectors was a sculptor named Pompeo Leoni from Sardinia. In the 1580s Leoni bought most of what the monk had saved, plus many other notebooks, breaking them up, cutting up some pages, and reorganizing them to his liking. I suspect that thousands of pages of irreplaceable writing were simply burned, because another two hundred years would pass before interest was taken in Leonardo's thoughts, science, and writings. Most of Leoni's collection resides today at Windsor Palace. Ironically, his acquisitiveness and vandalism of these precious works, simply by keeping them together, has unintentionally contributed to much of the last 150 years of great scholarship about Leonardo.

When Napoleon conquered most of Italy in 1796–99, he carted away to Paris as war booty all thirteen of the Leonardo notebooks that were in Milan's great Ambrosian Library, along with much other Italian art and treasure. These notebooks were the largest cache of Leonardo's writings and sketches still together following Melzi's death. After the duke of Wellington defeated Napoleon at the Battle of Waterloo in 1815, he ordered the notebooks returned to Milan. Perhaps the French misunderstood, because they only returned one (admittedly the relatively large, fabulous, twelve-volume Codex Atlanticus); the other twelve manuscripts remain in the Institut de France in Paris to this day. Decades of diligent scholarly analysis, comparisons, and research have been spent trying to put Humpty-Dumpty together again, particularly since the first publications of writings from the notebooks in 1883 and 1938. Still, we would not recognize much of it as linear. In large measure, this "failure" is due to the fact that they were never linear in the first place, in Leonardo's own writing. The mess is reflective of the marvelous tangle of Leonardo's thoughts, both detailed and sweeping, about so many matters that define the man. To a large degree, we just have to live with it.

THE FISHING EXPERTS

THERE IS AN IMPORTANT GROUP of men and women in this world, better anglers than the expert amateurs but not legends of national or international renown—the guides. I have had the good fortune to fish with and without local guides in sixteen states and ten countries; fishing with a great guide adds an entire dimension to the experience. Leonardo would have been the best of them all.

My first four years of fly-fishing were always with a guide. Gradually I got more comfortable rigging up, tying my own knots, and occasionally selecting a fly from those I knew were most frequently used by the guides. But the constant advice on where to cast, when to keep trying and when to move to another spot, and what mistakes to correct was invaluable. I vividly remember that day in 1988 when I gave up my training wheels, nervously wondering if I could put it all together and actually catch a fish alone on a small Colorado river called the Florida ("Flohr-ee-duh") rushing and gurgling through a ponderosa pine forest with a few boulders in the middle. When I held and then released that first nine-inch brown trout, I was awfully proud. In the following hours, days, and decades, I learned how the perfection I thought was

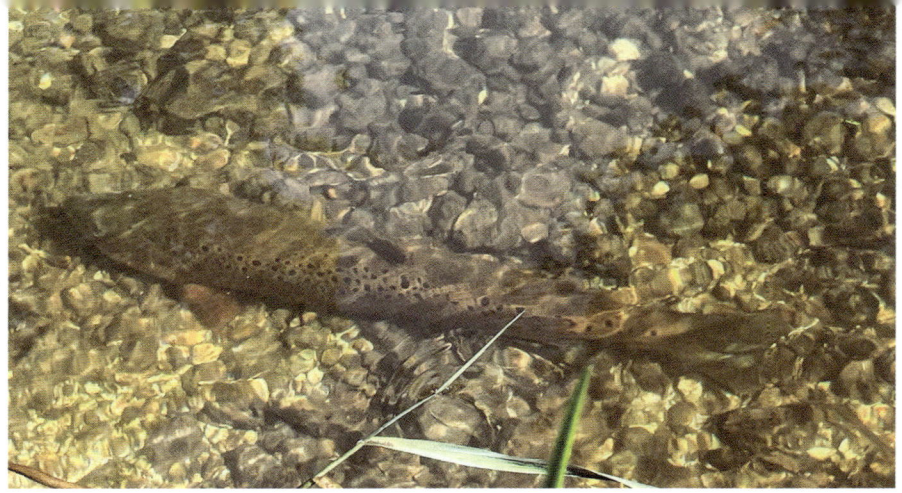

Can you spot the fish?

necessary in every step is not always necessary, how forgiving some fish in some rivers can be on some occasions.

Today I fish about 90 percent of the time without a guide, usually alone, usually successfully, whatever that means. But every time I fish with a good or great guide, I learn new things and catch more fish and have more laughs. I wish I could fish with Leonardo da Vinci as my guide.

Good guides see more fish. They know where to look, and they are much better at picking out small movements and shadows and the shape of trout, which are amazingly well camouflaged against the riverbed, covered by dappled light on moving water.

I have had a recurring conversation with Nick, guiding me on rivers large and small, in bright sunlight and on cold, overcast evenings.

"There's a big rainbow! See him?" Nick says.

"No. Where?" I say.

"Right there by the red rock, just twenty feet upstream and seven feet out from that bush!"

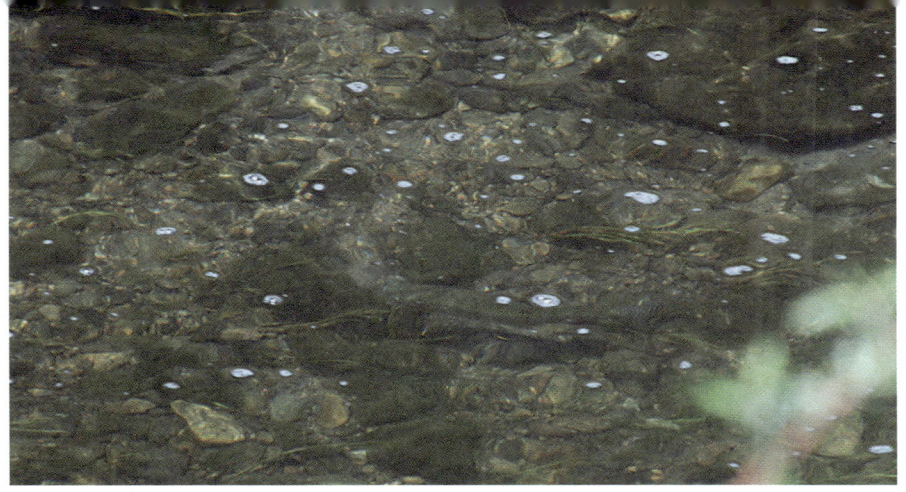

It takes a guide's eyes to spot this fine, camouflaged rainbow trout
(just below center, facing left; one small and one large spot
of white foam are over its back).

"What rock?"

"It's about three feet to the right of the dark rock."

"Okay, now I see the rock, but I don't see the fish."

"Hey! There's a bigger fish that just moved in to the right of the first one; see 'em now?"

"No, I still can't."

"Damn! We spooked them. I shouldn't have pointed at him."

"Yeah, I saw one fish when it swam off, but I never saw the other one."

Good guides add perspective. When have you left good water untouched? When have you flogged it to death?

Good guides read the water better. What spots should you walk on past? How close can you get to the possible lie? Where is the perfect place to stand to avoid drag and have the best chance to land a fish you have hooked?

Good guides tell you what you did wrong, with kindness. Good guides never yell.

Good guides work hard. They try different combinations of flies, weight, distance, and tactics until something works. They think outside the box when nothing is working.

Good guides are good company. They know when to teach, when to talk, and when to listen. They know when to joke and when not to. They thread the needle of controversial topics with care and grace.

Good guides are good guides. They know where to take you and how to get back. They care about safety and only wear you out to a safe degree. They don't quit before you're done, but they also don't make you feel bad when you want to quit early.

Good guides never touch the rod. They only fish or demonstrate when you ask them to.

Good guides are resourceful. When the keys are locked in the car (it is still possible) miles from the pavement, they know how to use fishing rods as break-in tools.

Good guides care. They want you to catch fish, to improve your skills, and to have a good time.

BEARS, DELUGE, AND EXPLODING MOUNTAINS

"TO STEP INTO A RIVER IS TO BECOME PART OF NATURE," writes Mark Kurlansky. I could not agree more, because the flow of the river, the "impetus," Leonardo called it, is primal. The gravity, the water, the current were all here before we came and will be here long after we leave the river, for the day or forever. To be part of it, or at least to be pushed around by it, feels special even without the excitement of catching a fish. To be attached to a trout by hook and line, feeling the head shake, feeling it go in unpredictable directions, is to feel pure wildness. To hold this beautiful, living, wild animal in your hand for a few moments if you get it into the net is as exciting to me today as it was forty years ago. It is literally nature incarnate.

But there is more. I take the same stream lunch every day that I fish. A small ziplock bag filled with low-weight, low-bulk, high-calorie food is stuffed in my fanny pack with my flies and rain gear, my water bottle and leaders and stream thermometer and other flotsam and jetsam that I might, but probably won't, need. My lunch is always salami and two kinds of cheese and three kinds of dried fruit and nuts and seeds of some sort. And as much as lunchtime provides nourishment when I force myself to stop fishing for ten or fifteen minutes on the stream bank, it often provides some of the most remarkable moments of the day. Because with my stream lunch, I look up and see things nearby, things I miss while my eyes are glued to the water.

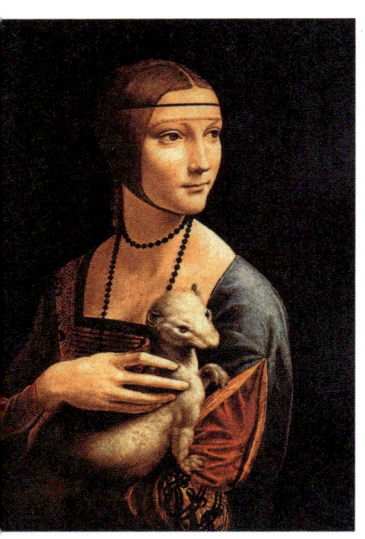

Detail from *Lady with an Ermine,* attributed to Leonardo, 1483—84

I see deer and squirrels and hawks, of course. But occasionally I see a bald eagle or an elk. Twice I've seen a mink, swimming or walking with the curious but stealthy gait of a caterpillar, fur shining brilliantly from the sun filtering through the leaves down to where it moves along the ground. Once, only once, I saw a pine marten just across the narrow river from where I sat, stock-still. It was so sleek and pure, it could have been Cecilia Gallerani's ermine. Once I heard a large rainbow trout hen literally smacking her lips as she hoovered up midges in an eddy, returning again and again, counterclockwise against the clockwise current, in a small crescent bight in the bank.

And I see bears. Usually I see bears running away. They smell me or hear me before I see them, and they have no use for me. They use their precious energy, gathered at something like ten thousand calories a day in the autumn, to put distance between themselves and my human scent and bother. It takes a lot of ants and acorns to make ten thousand calories; running away must seem like a lot of wasted work if they do the math of an hourly wage.

I have gotten used to bears. When I started fishing in New Mexico and Colorado, I was worried about them, those sharp claws and powerful jaws.

One morning I was struggling through a hundred yards of willows to reach the Rio Chama with my late friend Jim, a former veterinarian and lifelong resident with enormous knowledge about the area. I complained to him that this looked like an awfully likely place to surprise a bear up close and personal if one was upwind of us. Jim said, "You should count yourself lucky if you see a bear." I never forgot that advice, and I follow it.

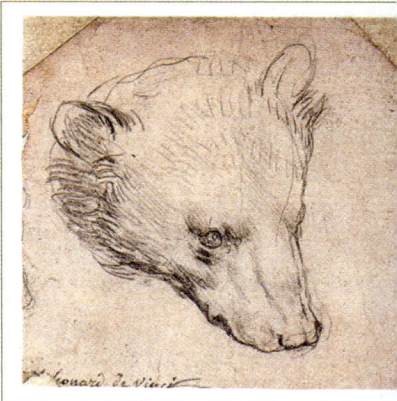

Bears walk by our cabin on their way between the high country and the river. Last summer we saw fourteen of them. Although they are black bears, they are seldom black. Sometimes they are brown; sometimes they are blonde, a sort of honey color. Sometimes they are mottled and scruffy. But usually they are magnificently shiny in their thick coats of fur, which probably are more useful to resist the scratchy scrub oaks than to keep them warm. Once, from our cabin, we spotted a bear that had dark legs and paws and areas around its eyes but was otherwise blonde; it looked like a giant panda,

Head of a bear

———

although I'm not aware of any of those running loose in North America. A few times young bears have come up to the screen door of our kitchen when Claudia is cooking a fragrant tomatillo soup; she calmly shuts the door and they wander away.

One fine afternoon I unhooked another wriggling, six-inch brook trout on a favorite small creek, which is only about ten feet across, in a steep and narrow canyon. I say "another trout" because I'd been catching fish on dries all day. This creek is filled with little rainbows and brookies, and twice I've caught more than one hundred in a day. I tell friends that no other guide in the world will guarantee they'll catch a fish if they just put a fly on the water. It's a bit of a trick promise, though, because it is not that easy to put a fly on the water there: the willows hang low over the creek everywhere and catch most flies in midcast. I fish there with a little six-foot rod and had to learn the bow-and-arrow cast to have a chance.

After tossing that pretty little trout back in the stream (they are much more resilient than larger fish, which have to be carefully released only after they get their "breath" and energy and balance back), I brought the tiny caddis fly up to my face to inspect it, dry it off, and see if it needed to be regreased with fly floatant. I was staring carefully at the fly through magnifiers, holding it about eight inches from my nose.

Something caused me to shift my focus away, to the bank, where a bear was standing up about fifteen feet from me. When I say standing, I mean on his hind legs, leaning against a tree, watching me. To this day, the image in my mind is of a man leaning against a lamppost with one foot crossed in front of the other, watching me for amusement. I'm not sure if that bear was thinking of taking up fly-fishing, or wondering why I was so dumb as to throw a perfectly good trout back in the creek, or wondering if anglers are good for dinner, but I knew he was too close.

I yelled at him: "Go away, bear! Go AWAY!" He leisurely lowered himself to all fours and walked into the bushes about five yards downstream, not acting the least bit scared of me, like he was supposed to be. I lost sight of him— not good—so I kept up my yelling. Then he waded out into the stream, and I figured he had just been trying to get from the right bank to the left, to make his way home. I relaxed. But when he got to the other side, he started walking back up that bank toward me. This was also not good. There really was no place for me to go. So, I picked up a rock and threw it toward him and yelled again. He stopped but did not run. I threw another rock and then one into the river to make a loud splash.

He turned around and sauntered back across the creek and into the bushes. I couldn't see him, didn't know if he was coming toward or away from me or going back up the right bank where I would happily see his rear end. Now I was sort of shaking. I waited and waited with another rock in my hand, but still he did not appear. After a few minutes, I remembered that I had an emergency bear whistle in my fanny pack to make a loud noise to scare bears away in just the situation I'd been in ten minutes ago. It took me several more minutes to find it among all the junk in the pockets.

I decided that it was an excellent time to call it a day and get back to my car, half a mile downstream. But the last place I had seen the bear was precisely that direction. As I walked, waded, and stumbled downstream, I blew my whistle and banged my wading staff on every rock when I didn't have a clear view for at least twenty yards, which was most of the time. I never saw that bear again, but I suspect he and his bear buddies had a good laugh around the campfire that evening, talking about the hapless fisherman he'd scared the wits out of that day.

One of Leonardo's most difficult drawings for me to understand is of an exploding mountain. It is not as if we have to tease out what it is, from a scene that we cannot imagine; he explained it clearly in an instruction to himself and to the students he imagined would read his "Treatise on Painting":

And let the fragments of some of the mountains descend into the depth of some valley and there form a bank to the swollen waters of its river; which having already burst its banks is rushing on with enormous waves, striking and destroying the walls of the cities and farms of the valley. And let the ruins of the high buildings of these cities throw up much dust rising up in the shapes of smoke or wreathing clouds against the descending rain. And as great masses of debris falling from the high mountains or large buildings strike the waters

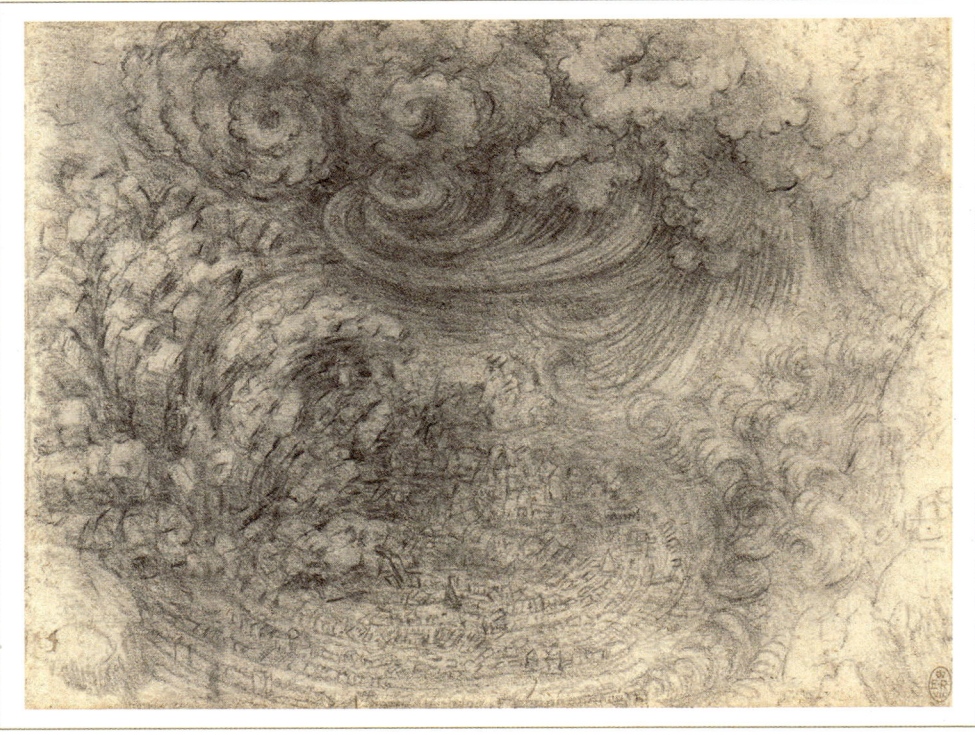

Leonardo, *Deluge*, 1517–19. Towering rocks on the left
fall on the town in the lower center.

But was Leonardo describing something he imagined engineering, as a massively destructive military tactic? Or something apocalyptic, as his worldview darkened near the end of his life? Or something else entirely? Bear in mind that this was one of his finished drawings, not one of the thousands of small sketches in his notebooks. Have you ever seen a mountain exploding from water? I have.

One of the first places a fly angler is taught to probe as a probable lie for trout is downstream from a rock or boulder. It is the ideal spot for resting in a slow current while the faster current brings a smorgasbord of food and calories just across the seam around the rock. Leonardo repeatedly observed and beautifully drew the increasingly complex seams and pools that accompany obstructions. The larger the rock, the more complex these currents are shown, including turbulence, eddies, foam, and spray.

But how did those boulders get into the river? As beautiful as they may be, they were not placed by a landscape architect with a bulldozer. They either fell into the river where they rest today or somewhere upstream, or they were pushed there by a glacier.

One day Nick and I were fishing a huge box not far from my cabin. In New Mexico, a box is a canyon that is pretty much blocked at the top by a slope or wall or waterfall that is too steep to climb, at least if you value your bones and your life. My friend Bill, a retired banker, calls this particular box not just the prettiest place he's ever fished but the prettiest place he has ever seen. It is not an idyllic stream winding through a green meadow but a dramatic and ominous canyon with cliffs rising two thousand feet on either side.

The water is crystal clear, which can deceive you into wading in too deep. And wade you must, because one side of the river or the other is nearly always that high, solid wall of schist. As we fish our way upstream, we cross back and forth to small areas with, if not exactly a flat place to stand, at least some huge boulder we think we can scale.

This river originates in natural alpine meadows at an elevation of nearly eleven thousand feet. Down in the box, it is wonderfully lonely due to both New Mexico custom and the fact that it is physically impossible to access except via the lodge that owns both sides of the river, followed by miles of hiking. I always thank the owner and his family for allowing lodge guests to fish there; it is just the type of place that is increasingly turned into totally private ranches in the western United States. Remarkably, the owner does not fish at all but enjoys the place instead by tirelessly hiking its steeps with his own friends.

Each time I enter the box where the river makes a sharp left turn, I feel both exhilarated by the beauty and the possibility of good fishing ahead and nervous because of the danger. The obvious danger is that the canyon twists and turns and you can only ever see about one-eighth of a mile upstream before a cliff wall obstructs the view. This adds to the mystery of just how far it is to the possibly mythical waterfall (everyone has heard of it, but only a few of us have reached it). I call each corner of the soaring cliff a "gate"—you cannot see around it until you reach it. When you pass the gate, a new beautiful vista opens up, but it too is blocked off by the next gate. My nervousness comes from knowing that no matter how sunny the sky above me may be, if it is raining hard in those broad meadows on top, there will be a flash flood down in the steep, narrow canyon where I am fishing.

As I said, those big, trout-sheltering boulders in the river came from somewhere, and, looking up at the cliffs, it is not hard to figure out where. We pick our path carefully around and over them, sometimes traversing recent rockslides that may or may not be stable. Although each of the hundreds of rocks apparently locked together in a slide probably weighs more than I do, occasionally one moves when I step on it, raising the specter of the whole thing sliding downslope a foot or twenty, with no concern about my bones.

Do you remember learning about erosion (which Leonardo described three hundred years before Hutton and Lyell) when you were in sixth or seventh grade? How mountains and rocks are ground down by wind and water and other rocks into gravel and soil? Remember the part about water seeping into cracks and then freezing and expanding into ice and making the cracks wider and wider with the passing years, until finally a rock breaks loose? It happens just like Ernest Hemingway's character Mike, in *The Sun Also Rises*, described how he went bankrupt: "Gradually, then suddenly."

The noise was not all that loud when it fell that day, but the splash was epic. The boulder wasn't even that large, maybe three hundred pounds, maybe twice that. But the significance was obvious to us: if Nick had been standing twenty-five yards downstream, where he'd been standing ten minutes earlier, I would be calling him "my late fishing buddy and guide." The impact of such a mass after falling for two seconds would be decidedly fatal, whether it hit you on the head or anywhere else.

So mountains do explode. Leonardo is with us when we fish. Be careful. Fishing is not always as graceful as Brad Pitt casting in *A River Runs through It*.

SUCCESS OR FAILURE?

Pass through this brief patch of time in harmony with nature,
and come to your final resting place gracefully, just as a ripened olive
might drop, praising the earth that nourished it and grateful
to the tree that gave it growth.
— Marcus Aurelius

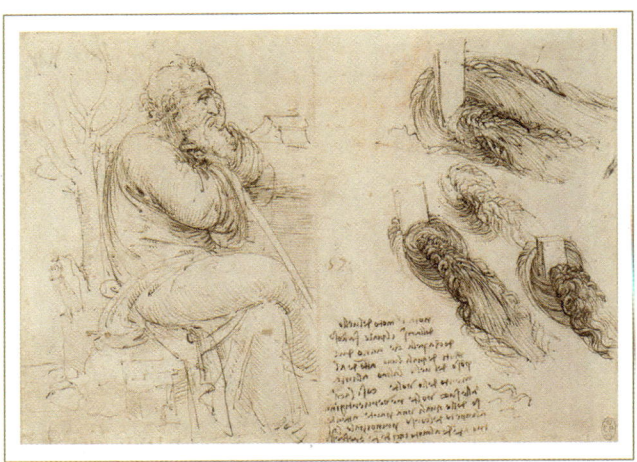

Old man and river currents

IN LATE 1515 FRANCIS I, the tall, twenty-one-year-old, newly crowned French king, met the sixty-three-year-old Leonardo. Francis began encouraging Leonardo to move to France. The king had established his court on the Loire River, near Blois, and planned to construct a new palace and capital. The following summer Leonardo made the nine-hundred-mile trek from Rome to Amboise—his first time outside Italy. Traveling this enormous distance by mule, horse, and chair, crossing the Apennines and then the Alps, must have been arduous; but it was well rewarded. Leonardo and his entourage were installed in a beautiful manor house just a ten-minute walk from the king's chateau and given stipends with virtually no obligations. Leonardo was appointed first painter, engineer, and architect to the king.

In some ways Leonardo was just an ornament of the court, but he filled the role admirably and happily, wowing visitors to the king's table and staging pageants to entertain them. He continued drawing, designing buildings and fountains as part of a planned palace complex and ideal city at Romorantin to the east, and sketching canals between the Loire and nearby Saône.

The young king's mother had envisioned Leonardo bringing the culturally advanced Italian Renaissance to France along with his wisdom. Francis I was deeply curious, intellectually capable, and "completely enamored" with Leonardo, the man of experience and knowledge. The mentoring relationship was made in heaven. Leonardo's long quest to replace the stigmas of illegitimacy, homosexuality, and procrastination with respect, fame, and financial comfort was fulfilled at last.

In the end, I am left with an unanswerable question: Would Leonardo enjoy fishing? I know he would have been truly great at guiding, for the many reasons already stated. But would he enjoy it? I believe he would have. The unity of nature. An appreciation of how perfectly suited to their watery environment his creator had designed fish to be. The appeal to all the senses and the beauty of natural surroundings. The unique pleasure of holding for a moment such a wondrous, wild creature in his hand. And then releasing it to swim away.

He would appreciate that these small fish survive floods that cannot be resisted by the creations of humankind. He would notice the tangible evidence all around him of the two faces of rivers he described: creation and destruction of nature. He tried to harness the rivers for both good and war, which he detested.

Did Leonardo succeed in his life?

His plan to divert the Arno was audacious, crazy, and impossible, just as building the Panama Canal was audacious, crazy, and impossible until it succeeded. But his plan never came to fruition. His monumental equestrian sculpture for the duke of Milan was never finished. Virtually all of his commissions in Florence were left unfinished. He never achieved financial security without a generous patron. Not one of his treatises was published in his lifetime. It sounds like failure.

And yet consider the hold that Leonardo, the scientist and painter, has on us more than five hundred years after he joined the ranks of dead white men.

Ten million people visit the *Mona Lisa* every year, and it is certainly the most famous painting in the world.

Vitruvian Man is the world's most famous drawing and is used to promote countless products and ideas. Young people often think the man in the drawing is Leonardo.

In 1903 the Wright brothers proved Leonardo's theory that humanity could develop a way to fly and confirmed his belief that the first ones to do so would achieve everlasting fame.

In 1994 Bill Gates purchased the Codex Leicester for the highest inflation-adjusted price ever paid for a book or manuscript. Today this notebook about water and moonlight is still considered the world's most valuable.

In 2000 Adrian Nicholas jumped off a balloon ten thousand feet above South Africa to test Leonardo's parachute design, built only with materials available to Leonardo. It worked. Martin Kemp quotes the intrepid Nicholas: "It took one of the greatest minds who ever lived to design it, but it took five hundred years to find a man with a brain small enough to actually go and fly it."

In 2002 aviator Judy Leden proved the viability of Leonardo's machine for human flight in a batwing glider constructed per his design with materials that would have been available to him. Unpowered, she flew farther than the Wright brothers' first flight!

In 2006 Morteza Gharib of the California Institute of Technology completed and exhibited a model of the human aorta and its valve, demonstrating Leonardo's ingenious deduction that vortices in the flowing blood are what cause the valve to work. Only in 2014 was the proof made final, by real-time MRI observation of a living person at Oxford University.

In 2011 engineers at the German company Festo completed an artificial herring gull with a six-foot wingspan according to Leonardo's observations and drawings of how birds fly. It takes off by itself, flies around with other seagulls, and corroborates another one of his complex and unprecedented designs.

In 2017 a painting probably by Leonardo sold for the highest price ever paid for a work of art.

In 2021 a tiny drawing by Leonardo of the head of a bear sold for more than $12 million.

In 2022 aerospace engineering graduate students at the University of Maryland proved that Leonardo's 540-year-old "helicopter" design also worked. The aerial screw, when made out of modern, lightweight materials, produced enough lift to fly a quadcopter drone. To everyone's surprise, the screw design in motion also produced a vortex, such as Leonardo so consistently studied in water, that pushed down against the air below to help develop the lift needed to fly and with less downwash, an unanticipated bonus. (Particularly remarkable to me is that Leonardo's designs, done in secret, received no helpful feedback from colleagues.)

In 2023 scientists confirmed and finally explained Leonardo's astute observation that tiny air bubbles rise in water in a unique spiral pattern.

Also in 2023, Caltech professor Gharib and his colleagues published Leonardo's experiment notes and diagram regarding gravitational acceleration, which Morteza had found in 2017 in the Codex Arundel. Although Leonardo mistakenly used an incorrect exponent, Gharib and his colleagues showed that he had a mathematical knowledge of gravity more than 75 years before Galileo and about 175 years before Isaac Newton. Leonardo was off by only 3 percent in calculating the gravitational constant.

On one hand is the criticism of Leonardo for being an underachiever because he completed so few paintings and no sculptures. His greatest sculpture project was abandoned for lack of bronze (which was diverted in wartime for making cannons), and what was left of the mold for the giant horse statue in Milan was used by invading French soldiers for target practice. Even Leonardo asked in a notebook on a discouraged day, "Was anything ever done?"

On the other hand, the world's most celebrated painting and drawing are the products of his hand. His works were desired by nobility in his lifetime and by billionaires today. His constant presence at court was sought by the most powerful king of his time. His scientific treatises, never published, are still celebrated, and every exhibition of his art or scientific work around the world brings crowds and lines. Nearly all people who know little about his life describe him as a genius. I certainly believe he was a success and considered himself to be one. How could he not be?

When Leonardo died, he did not believe that an immortal soul could live apart from its physical body. But if not in heaven, then certainly on earth, with all of us who love and admire him, who choose to learn from him, Leonardo lives on.

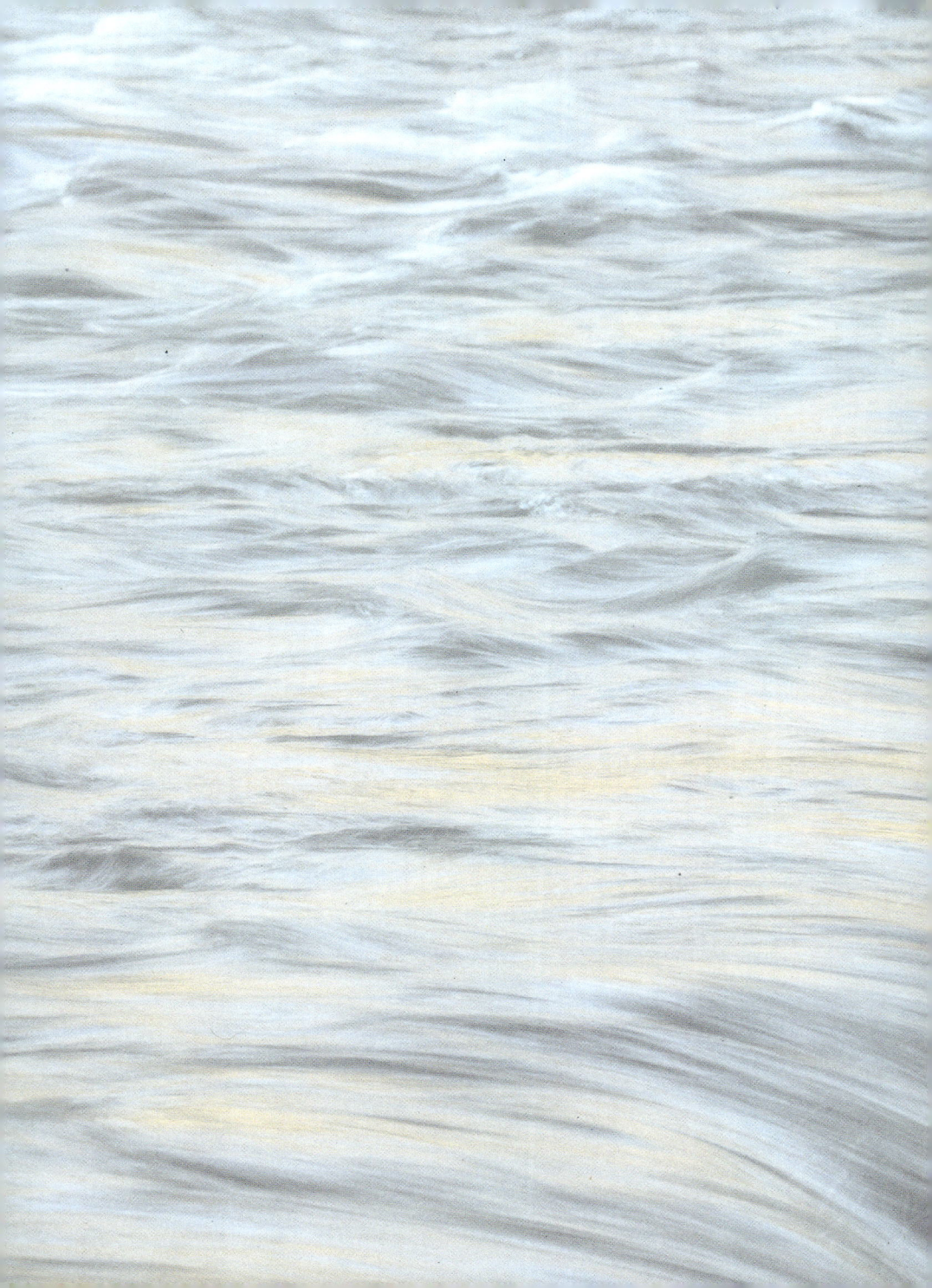

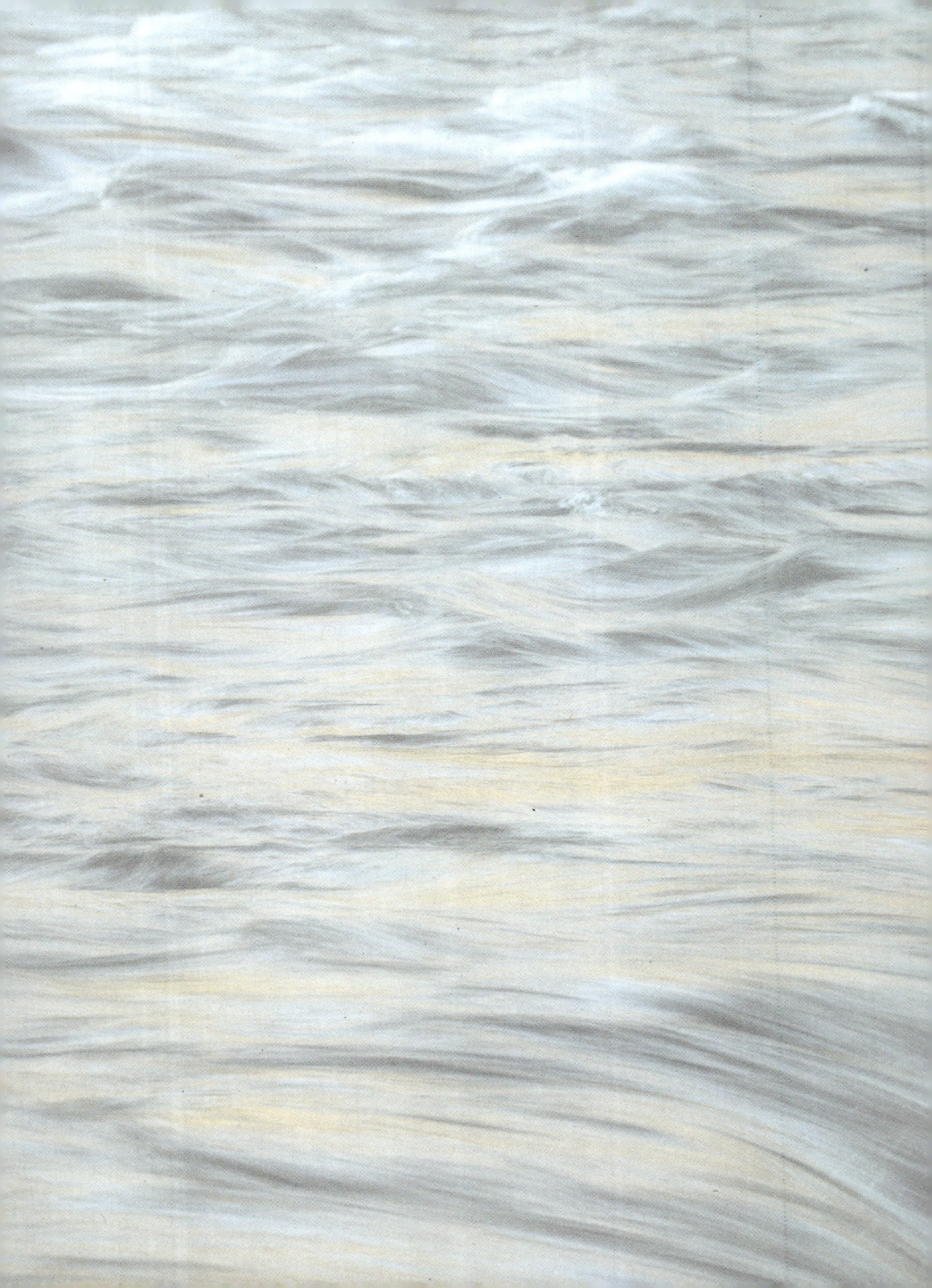

ACKNOWLEDGMENTS

I ORIGINALLY SET OUT MERELY TO INTRODUCE ANGLERS to Leonardo's drawings of water currents. The sustained encouragement of others has been the most important factor in bringing this book to fruition as it is. Claudia Ladensohn, my wife of fifty years, never asked me to bring Leonardo into our marriage, where I have injected him into every discussion of every topic, no matter how irrelevant or inappropriate. She was the first to read my drafts and has offered support from beginning to end, from looking at the entire project to debating punctuation. My daughter, Eliza, and her wife, Ariel, pushed me to write for a general audience and have been unflagging in their enthusiasm these many years.

My big sister and author, Sydney Ladensohn Stern, offered invaluable professional advice, perspective, and critical reading at inconvenient times. Repeated, skeptical questions from my middle sister, Ann Dee Steidel, about what I was really writing and for whom, truly forced me to decide.

Rick Cohen took me fly-fishing forty years ago and created a monster. He has listened patiently every week to my fishing stories and always believed that I could complete this project. He has innocently joined many fishing trips

I dreamt up; not all were well advised, but they all produced laughter and memories. Jan Cohen introduced me to Ken Burns and the next generation of documentary makers, Sarah Burns and Dave McMahon, whose parallel path has energized me. Cousin Chuck Katz is the only fly-fishing art historian I know; his immediate grasp of the project, along with his encouragement and that of his wife, Roberta Katz, were enormously helpful.

Martin Clayton and Martin Kemp each gave so generously of their time and expertise that I left England inspired and committed to doing my very best to justify their help. Other scholars answered numerous questions for me, including Julianna Barone of University of London, Leslie Giddens of Tulane University, Roald Hoffmann of Cornell University, and Sara Taglialagamba of University of Urbino Carlo Bo and Nuova Fondazione Rossana e Carlo Pedretti. This a work of nonfiction; I have tried my best to stick to the facts and label my speculations as just that. The historical characters are described as I see them, and whatever mistakes I have made are solely my responsibility.

Four friend-scholar-authors have repeatedly provided me with excellent advice, perspective, and encouragement: Char Miller, Michael Nye, Katherine O'Rourke, and Naomi Nye. My special gratitude to each of you.

Andrea Clarke and Kathleen Doyle at the British Library, "Jet" Jacobs and her team at the University of California Los Angeles Library Special Collections, and Karen Lawson and Daniel Partridge at the Royal Collection Trust could not have been more helpful and cheerful in guiding this pedestrian through their rarified and fascinating worlds. I received particularly valuable and timely advice from Lewis Fisher, Catherine Atherton, Tim Garson, Bill Reid, Ed Cross, Wendy Atwell, Bob Kolker, Judi Lipsett, Andrew Klotz, Hannah Beresford, and Josh and Eliana Armstrong.

Anna Pozzi, Eros Villa, and Marcella Vimurcati gave a uniquely welcoming and complete tour of the Vaprio d'Adda region where Leonardo lived and worked in 1512–13.

Taylor Streit taught me how and where to fish my home waters and about publishing too. Nick Streit teaches me new techniques, new waters, and new thinking every time I am with him. His advice and his photography have made him a true partner in this book.

Three other generations must be mentioned. My short-tempered grandfather personified patience and unconditional love toward me through years of teaching me to fish as a child. I, in turn, had the privilege to teach my own father to fly-fish late in his life. And my son, Graham, loved to fish; I treasure remembering him winning three rods and reels in one day at a local fair.

Trinity University Press's Tom Payton evinced a vision and enthusiasm for the project from day one, backed up with a team of professionals: Sarah Nawrocki, Anne Boston, Daniel Simon, Emily Schuster, Burgin Streetman, Bridget McGregor, and others.

I am especially grateful to you, the reader, for listening to this story.

Finally, I must acknowledge Leonardo himself. I never met the man, but he has introduced me to many of the people named above, whom I respect and enjoy. He has inspired me with his curiosity and rigor, made me think differently, and enriched my life. Thank you, Leonardo.

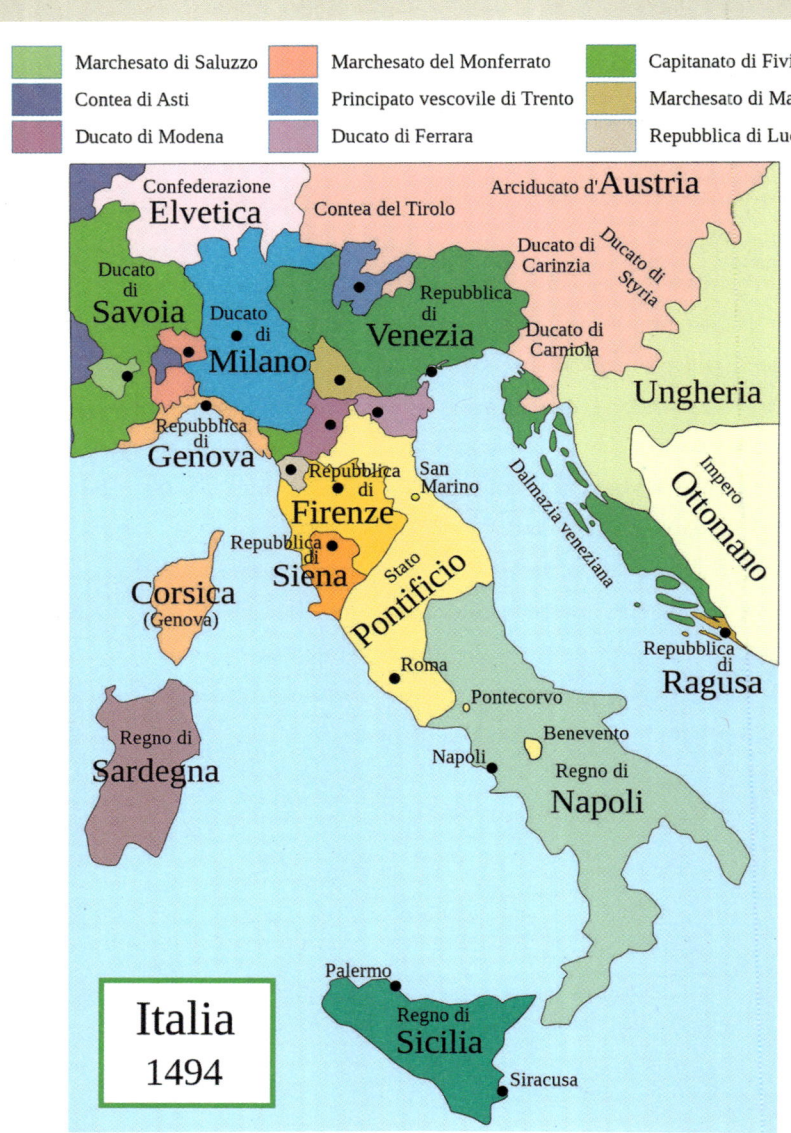

Legend

Marchesato di Saluzzo	Marchesato del Monferrato	Capitanato di Fivizzano
Contea di Asti	Principato vescovile di Trento	Marchesato di Mantova
Ducato di Modena	Ducato di Ferrara	Repubblica di Lucca

Confederazione Elvetica

Contea del Tirolo

Arciducato d'Austria

Ducato di Savoia

Ducato di Milano

Ducato di Carinzia

Ducato di Styria

Repubblica di Venezia

Ducato di Carniola

Ungheria

Repubblica di Genova

Repubblica di Firenze

San Marino

Dalmazia veneziana

Impero Ottomano

Repubblica di Siena

Corsica (Genova)

Stato Pontificio

Repubblica di Ragusa

Regno di Sardegna

Roma

Pontecorvo

Benevento

Napoli

Regno di Napoli

Palermo

Regno di Sicilia

Siracusa

Italia 1494

Florence (Firenze) in 1494

TIME LINE

1054	Schism between Roman Catholic and Eastern Orthodox churches	
1066	French Normans conquer England	
1096–1192	Crusades to recapture Holy Land for Christendom	
1265–1321	Dante Alighieri in Florence	
1397	Medici Bank est. in Florence	
1343–1400	Geoffrey Chaucer in England	
1434	Cosimo de' Medici rises to power in Florence, supporting the arts and making Florence the cradle of the Renaissance	
1439	Delegation of 700 Byzantine church leaders, including the emperor, travel from Constantinople to Florence to meet with Roman Catholic leaders to heal the Great Schism	
1452		**Leonardo is born in Vinci, outside Florence**

1453	End of Hundred Years' War in Europe	
1466		**Leonardo apprenticed to Verrocchio**
1471	Albrecht Dürer born in Germany; Holy Bible translated into Italian	
1473		**Leonardo makes his first painting in oils and sketches "first ever European landscape"**
1470–75		**Leonardo completes Verrocchio's *Baptism of Christ* with extraordinary portrayal of water**
1475	Michelangelo Buonarroti is born in Caprese, Republic of Florence	
1478		**Leonardo's friend Berlinghieri publishes first modern version of Ptolemy's ancient atlas**
1483	Titian is born in Italy	**Leonardo has moved to Milan**
1490		**Leonardo imagines writing a book to teach painting to artists; 8,000-plus pages are never finished, 4,100 survive**
1491	Henry VIII is born in England	
1492	Columbus makes landfall in America	**Leonardo prospers in Milan**
1494	Michelangelo finishes sculpting *David*	
1496	Dame Juliana Berners, OSB (b. 1388, fl. 1450), *Treatyse of Fysshynge with an Angle*	
1499		**Leonardo leaves Milan as French invasion succeeds**
1502–5	Amerigo Vespucci's letters reveal discovery of a New World	**Leonardo back in Florence; tries to divert the Arno; Leonardo works on the "Codex Leicester" in Milan**

1512	Michelangelo completes Sistine Chapel ceiling	
1512–13		**Leonardo and company in Vaprio d'Adda, avoiding French-Sforza conflict in Milan**
1513		**Leonardo moves to Rome, taking the *Mona Lisa* with him**
1514		**Leonardo works on hydraulics for Rome**
1517	Martin Luther nails his 95 Theses to the door of Wittenberg Castle church and sparks Protestant Reformation	**Leonardo moves to Cloux at the behest of King Francis I of France**
1519		**Leonardo dies in Cloux**
1532	Machiavelli publishes *The Prince* in Florence	
1543	Copernicus publishes his Heliocentric Theory	
1558	Elizabeth I is crowned queen upon death of Henry VIII	
1564	William Shakespeare is born in England; Galileo Galilei is born in Pisa, in the Duchy of Florence	
1642	Galileo dies	
1653	Isaak Walton's (1593–1683) *The Compleat Angler* published in England; expanded and completed by Charles Cotton in 1676	
1776–99	American and French Revolutions	**Graveyard in France with Leonardo's remains is destroyed**
1848–71	Italy is united for the first time as a nation	

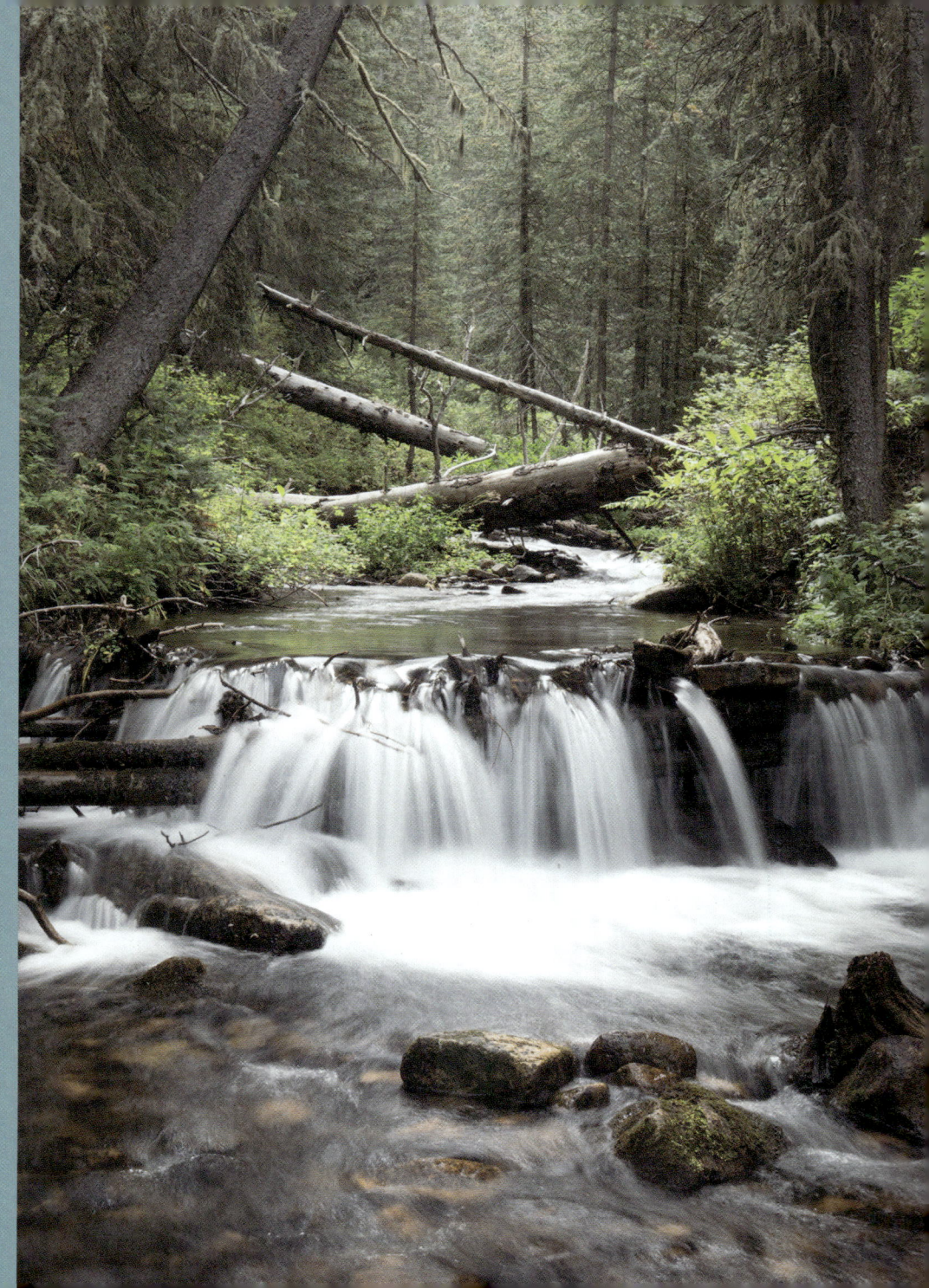

A FLY-FISHING JARGON PRIMER

AS IN ANY SPORT, PROFESSION, OR ORGANIZATION, fly-fishing has a lot of jargon. It's probably some anthropological signal of belonging, but that is above my pay grade. In case you are not already immersed in this jargon, here are some commonly used terms and approximately what they mean in vernacular English.

Rods resemble the ones used in other kinds of fishing but are different in several ways. Rods (also jokingly called poles, sticks, or wands) come mainly in graphite; rods made of bamboo (expensive) and fiberglass (pregraphite) are traditional-fishing cult varieties. Fly rods are usually eight to nine feet long, good lengths for developing line speed for casting and picking line up quickly off the surface to set the hook when a fish bites. For many different specialized purposes, rods vary from six to eleven feet or longer.

Fly rods come in different weights, from a very light 00-weight for tiny fish to 15-weight for landing sailfish and marlin. Most people fish for most trout with a 4-, 5-, or 6-weight rod. The variations in length, weight, action, speed, flex depth, and so on result in nearly infinite combinations for different fishing purposes, real and imagined. This allows fly-fishers to spend obscene

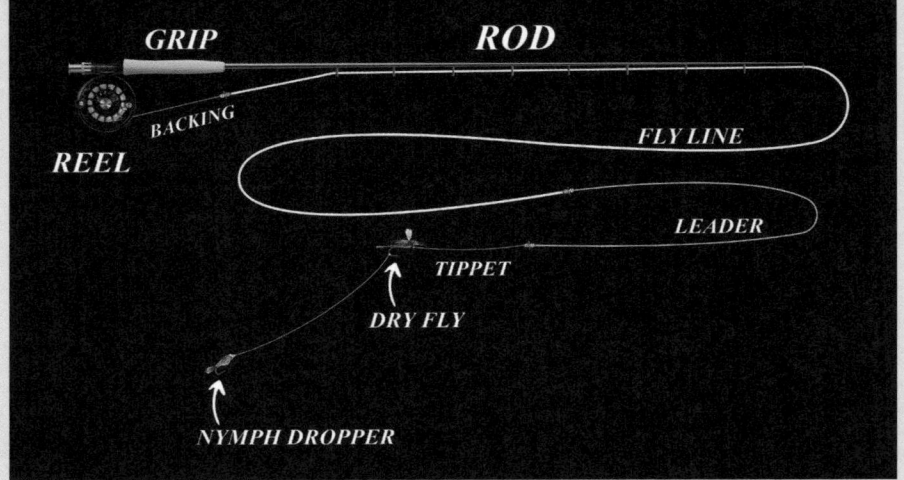

Fly-fishing rig

---

amounts of money on many rods when one or two would usually be plenty. It also prompts the old joke, "I hope that when I'm dead, my wife doesn't sell my rods for what I told her I paid for them." Other than casting, the fly rod's main purposes are to serve as a flexible shock absorber to keep a hooked fish from breaking the line and as a lever for bringing a strong or heavy fish to the net.

The bottom line with all of these combinations is that instead of getting "a new fishing rod," fly-fishers today might say they got "a nine-foot, 5-weight, fast-action, midflex rod," but they are still thinking about buying "a ten-foot, 3-weight, soft-tip rod for Euro nymphing."

Reels in fly-fishing are much less for pulling in fish than they are in other forms of fishing where reels have multiplying gears. Fly reels have a drag system that smoothly impedes the fish from easily taking out a lot of line, but cranking hard on a fly reel often means breaking the tippet and losing the fish. Most of the time, a reel is primarily a convenient place to store the fly line.

Tippet is intentionally the weakest part of the fly-fishing line. It is made of thin, clear nylon or fluorocarbon and intended to be invisible to the fish when the fish considers eating the artificial fly tied to it. The lighter and thinner the tippet, the less visible it is to the fish. The lighter and thinner it is, however, the more easily it will break. It connects the fly to the leader with various fishing knots that can withstand the force of a fish pulling against the entire rig. Tippet comes in diameters that shrink as the number goes up (the opposite of rod numbering, presumably to confuse the Russians): 2x and 3x are thick and strong; 6x and 7x seem like human hair and will break easily if the fish is too strong, or the reel's drag system is too tight, or the angler sets the hook with too much enthusiasm. Catching a big fish on fine tippet earns bragging rights because it means the angler did everything right (or got lucky and the fish was unusually cooperative).

Leaders connect the thick fly line to the thin tippet that the fly is tied to. Leaders taper from a thick end nearest the rod tip to a thin end near the fly and are usually seven to nine feet long. They are numbered by the thickness at the thinner end, and the numbers match the tippet numbers, like 4x and 5x. Leaders form a key part of the casting loop in the air and "turn the fly over" if the leader, tippet, and fly are approximately the right size for each other. (If mismatched, the last few feet of it all will collapse in a pile instead of laying out in a straight line on the water.)

Fly line is a plastic-coated fishing line approximately ninety feet long. It has a modest amount of bulk and weight in the first fifty feet but more weight than any other part of the fly-fishing tackle (excluding the rod and reel). Unlike fishing with bait or lures, which weigh something and are attached to light monofilament line, in fly-fishing the angler is casting the line itself, which carries the lighter leader, tippet, and fly with it out onto the water. Fly lines are numbered to coordinate with fly rods of that weight, so a 5-weight line is meant to optimally flex a 5-weight rod and, in turn, to be cast out when that particular weight rod unbends with a snap. It is the fly line that forms a loop in the air.

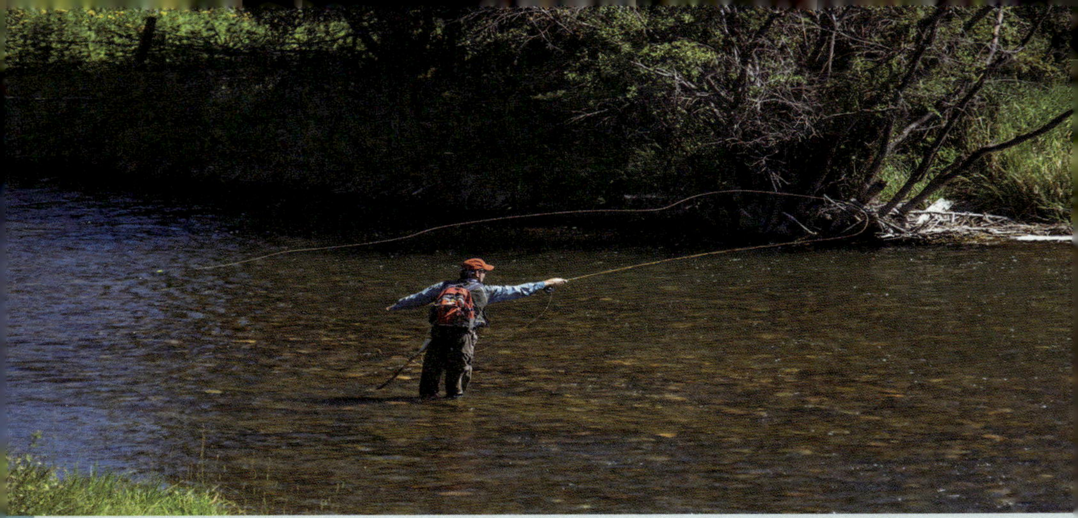

Casting a tight loop for accuracy and distance

The ironically scenic part of fly-fishing is an angler casting the fly line back and forth to develop the line speed and length to propel the line, leader, and fly to the targeted spot holding a fish. Fly lines come in different weights, tapers, and surface textures, but more importantly, most lines float on top of the water due to tiny air bubbles embedded in the plastic coating. For some specialized purposes like streamer fishing or lake fishing where fish are deep, a fly line can have lead or tungsten embedded in the entire length ("sinking line") or the ten to twenty feet closest to the fly ("sink-tip line").

If the number of rods, reels, lines, tippets, and leaders seems confusing, such abundance is dwarfed by the number of *flies*. Each fly pattern has a name, chosen by the person who invented it, and there are thousands of patterns. For example, a common fly that imitates adult caddis flies (sedges) is called an elk hair caddis.

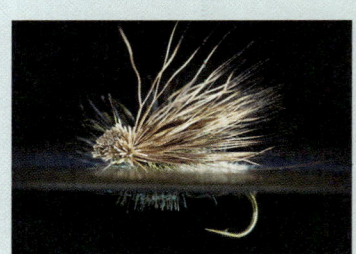

Elk hair caddis is tied to fool a fish into thinking it is an edible caddis fly.

That is the pattern, but it can be tied in many colors and sizes. Let's say it is commercially available in sizes 10, 12, 14, 16, and 18 (odd numbers are virtually never used, and higher numbers indicate smaller sizes); the body may be tied in gray, tan, brown, yellow, or olive colors. That means there are five sizes times five colors, or twenty-five different commercially available flies all called an elk hair caddis. The most common mayfly imitation, a parachute Adams, comes in at least ninety combinations. (These numbers, of course, don't include the variations that individuals tie for themselves.) Some patterns have less variety, and some have even more. The point is that with thousands of fly patterns, each in ten to a hundred varieties, there is something for every taste. Yet anglers never seem to have the exact fly they want with them to "match the hatch" of insects at a particular time. Some minimalists carry just a few flies of those patterns they have the greatest confidence in. Maximalists carry hundreds of patterns in multiple fly boxes or in a large accordion-style box on their chests. If you want to see photos of fly patterns and the insects they are meant to mimic, two good websites to visit are johnkreft.com/how-many-fly-patterns and hikingandfishing.com /types-of-fly-fishing-flies.

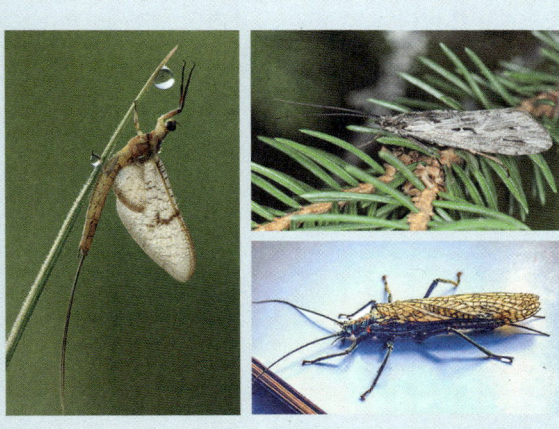

Mayfly adult (*left*); caddis fly adult (*upper right*); stonefly adult, approximate actual size (*lower right*)

Trout eat mainly aquatic insects, which live at least part of their lives in the water. The main groups (usually different orders of insect) are mayflies, caddis flies, stone flies, and midges. Each group includes many species (for example, there are six hundred mayfly species in the United States alone and three thousand worldwide), many sizes, and many colors, hence the huge variety of artificial flies we might use to fool trout into thinking our imitation is a real insect. All of these "bugs" lay their tiny fertilized eggs on the surface of the river or lake, where they sink to the bottom and stick to pebbles, sand, weeds, sticks, or gravel.

Eventually (usually within a few weeks), a small larval, or *nymph*, form of the insect hatches out of the egg and grows by filtering microscopic food from the water or eating bits of plants or even smaller creatures. As the nymph grows, it may shed its skin as many as fifty times over a period of months or even a couple of years. These underwater critters supply the main portion of a trout's diet, so we fish with imitation nymphs all year. Most are walking around on the bottom of the river or fastened onto something near the bottom, like a rock or a stick. So, nymph fishing involves getting the fly to sink down through the *column of water* between the surface and the riverbed and then drift downstream with the current, at the intended depth. Because individual insects are constantly being knocked loose from their homes by the rushing river, fish can find them high and low, but understanding the currents helps the angler imagine where most of the nymphs are likely to be. And that is where the trout are particularly likely to be.

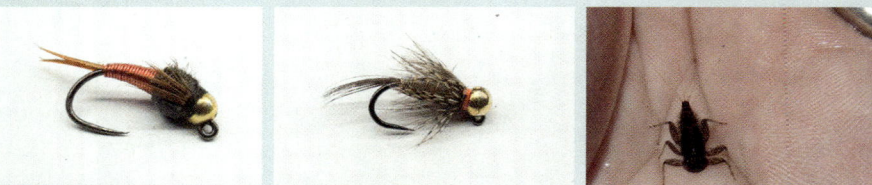

(*Left to right*) A Red Copper John fly; an HDA Favorite fly;
and a mayfly, greatly enlarged (note wedding ring for scale)

Spoiler alert: now comes the cruel part of the aquatic insect's life. The nymph is fully grown, ready for its exciting adult stage, including winged flight and wild sex. It somehow reads signals about the temperature of the air, where it has never lived; it shucks off its last nymphal skin and floats and swims, struggling to the surface. During these few seconds, we call the insect an *emerger;* it attracts trout attention as it moves up through the water column and frequently gets eaten in the process, often with an energetic take by the fish. But the signal is read by many individual bugs at the same time and there are lots more bugs than fish, so in fact most of these emergers successfully pop up on the surface. There they quickly dry out their shriveled, brand-new wings. But they cannot fly yet. During that minute or so on the surface (or trapped just below it by water's surface tension), these *duns* are drifting helplessly wherever the current takes them. The fish know where the currents are bringing a conveyor belt of newly hatched adults down the river; the *hatch* is another massacre of bugs by fish. Finally, the surviving adult gets lift-off. Mayflies shed one more skin to become *spinners*. Later, swarming in all its generation's glory above the water in a cloud of potential mates, the orgy of mating might last a few minutes or up to two hours. A few happy species make it as adults to, at most, a second day of mating. Fertilized females fall with their eggs almost immediately onto the water's surface, expelling them all at once, or they do repeated touch-and-goes, dropping eggs on the water each time.

Once again the trout have a field day, ambushing the adult as she touches down or floats along. The fly-fisher sees trout rising to the surface, making rings as they sip bugs from below, sticking their heads above water to open their jaws and chomp down, or coming clear out of the water to chase an adult insect leaving the surface. This is *dry fly-fishing*, using an imitation of an adult insect on top of the river's surface. Anglers love it because we can see the entire action play out. The visual experience Leonardo so privileged is often amplified by the sound of the trout breaking the surface, sometimes even snapping its jaws closed or smacking its "lips." To hear those sounds and turn around to see the ring a fish just made, gliding down the river, never ceases to make me smile. Since daylight and air temperatures are favorable for most insect species to emerge and mate in the summer, that is the main season for dry fly-fishing.

Finally, the exhausted insect falls to the surface, never to rise again. Mayfly wings in particular are outstretched, no longer folded over their backs. Often, such a *spinner fall* results in many "spent spinners" collecting in eddies, dead or dying, where trout happily hoover them up.

Most fishing flies are tied to imitate the appearance of a specific nymph (wet) stage or adult (dry) stage insect. We call them *imitators*. But there are also *attractor* flies, both nymphs and dries, that apparently just look generally buggy to fish. Although they do not imitate a particular insect, if the fish like them, we like them. To round out the jargon of flies, there are two more large groups that need mentioning.

First, there are nonaquatic insects that fall into the river and are good meals for fish. These are mainly grasshoppers, beetles, and ants. These *terrestrials* are always fished as dries and are particularly effective when fished along the edge of a stream, where *naturals* (real insects) would normally happen to fall in. Because they are usually larger than aquatic insects, these are particularly good patterns to drift on top of the water, with a nymph (or two in line, legal only in some states) tied on a piece of tippet below, sinking down into the deeper currents. If a trout eats the nymph *dropper* and the line is taut, the dry will be pulled underwater and the angler knows to set the hook. This arrangement of *dry-dropper* or *hopper-dropper* is extremely popular in the West, because it gives the fly-fisher a shot at fish feeding on both adult insects on top and nymphs down below. Any dry fly used this way is called an *indicator* because it can indicate that a fish has taken the nymph drifting below.

Finally, there are *streamers*. These are larger flies, fished below the surface but tied to imitate a minnow or small fish or leech. These naturals must have much more protein than a single insect; trout frequently attack this kind of fly with energy they do not spend on a small bug. There are more patterns still—mice, dragonflies, scuds, worms—but you have most of it now.

What are flies made of? The only common element is a *hook*, and there are, like the rest of fishing stuff, countless sizes, lengths, and shapes. The size of a fly actually refers to the size of the hook, specifically to the size of the gap between the shank and the sharp point. Most fly-fishers like to either use barbless hooks or mash down the barb to avoid tearing the fish's mouth when removing the hook. This also makes it easier for the fish to spit the hook and get away. The rest of the fly is made of thread, yarn, and bird feathers or animal hair. Not just any feathers but usually neck or tail feathers from a chicken, peacock, partridge, or pheasant. And not just any animal hair; usually fly-tiers use buoyant, hollow hair from elk or deer, or rabbit fur, or wool yarn, or plastic threads or pieces of tinsel, or whatever else the fly-tier thinks will make a trout eat it. Wet flies (nymphs) will, in addition, often be tied with wire or a heavy bead for a head, to help them sink.

If the fly-fisher decides to only fish subsurface nymphs, she can use a *strike indicator* to, well, indicate when a fish strikes the fly and it's time for the angler to *strike*, or set, the hook. This indicator serves exactly the same function as a bobber, but we don't like to use terminology like "pole" and "bobber," so we use jargon. Anyway, various types of strike indicators are designed to minimize friction with the water so that the nymphs are mainly moved by the subsurface currents alone instead of being pulled along by the indicator on the surface. Strike indicators are often spherical to minimize friction with the water or made from light nylon yarn. A sensitive strike indicator might twitch only a bit when a trout takes the fly gently and the angler must react quickly.

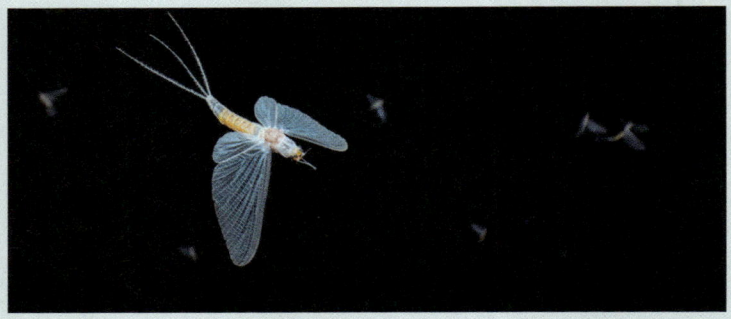

Adult mayfly in flight, "in her wedding dress"

Older versions don't move at all, and the fly-fisher never knows he's missed a chance to catch the wily trout.

Beginning in the early 1980s, fly-fishers in Europe popularized fishing with no dry fly, no strike indicator, and no floating line at all outside the reel. The objective was to have direct contact between the rod tip and the nymph with a tight line, feel the riverbed, and feel the fish's mouth touching the fly, telegraphed via a long, taut, thin leader and tippet. In competitive fishing tournaments, this approach produced more fish and therefore more championship national teams. At first it was called Czech nymphing (because Czech teams dominated tournament wins) and, more recently, Euro nymphing. This approach has caught on in the United States, where it is now often called *direct contact nymphing* or *tight line nymphing*.

CREDITS

Page 133, courtesy Roald Hoffmann

Pages 138 and 180, courtesy Baijou Abdeljabbar

Page 147, by the author

Page 152, courtesy Will Templer, Field Sports Services

Page 153, courtesy Ed Engle photo

Page 174, courtesy Wikimedia Commons

Page 182 (top), SSI-DUKE, iStock.com

Page 182 (bottom), courtesy @timromanophoto

Page 183 (left), Fabioski; (upper right), Henrik L, iStock.com; (lower right), courtesy Charles J. Katz Jr.

Page 188, courtesy Ferenc Kocsis

Pages 34–35, translation by Jean Paul Richter

Page 48, translation and paraphrase by Martin Kemp and Domenico Laurenza

Page 81, translation by Martin Kemp and Lucy Russell

Page 88, Codex Leicester, translation by Carlo Pedretti, quoted in Carmen Bambach

Pages 100–101, quoted, with permission, from Adrienne Mayor, *Flying Snakes and Griffin Claws: And Other Classical Myths, Historical Oddities, and Scientific Curiosities* (Princeton University Press, 2022)

Page 102, Dame Juliana's entire brief book is available for free at gutenberg.org/files/57943/57943-h/57943-h.htm

Page 128, translation by Emma L. Seeley

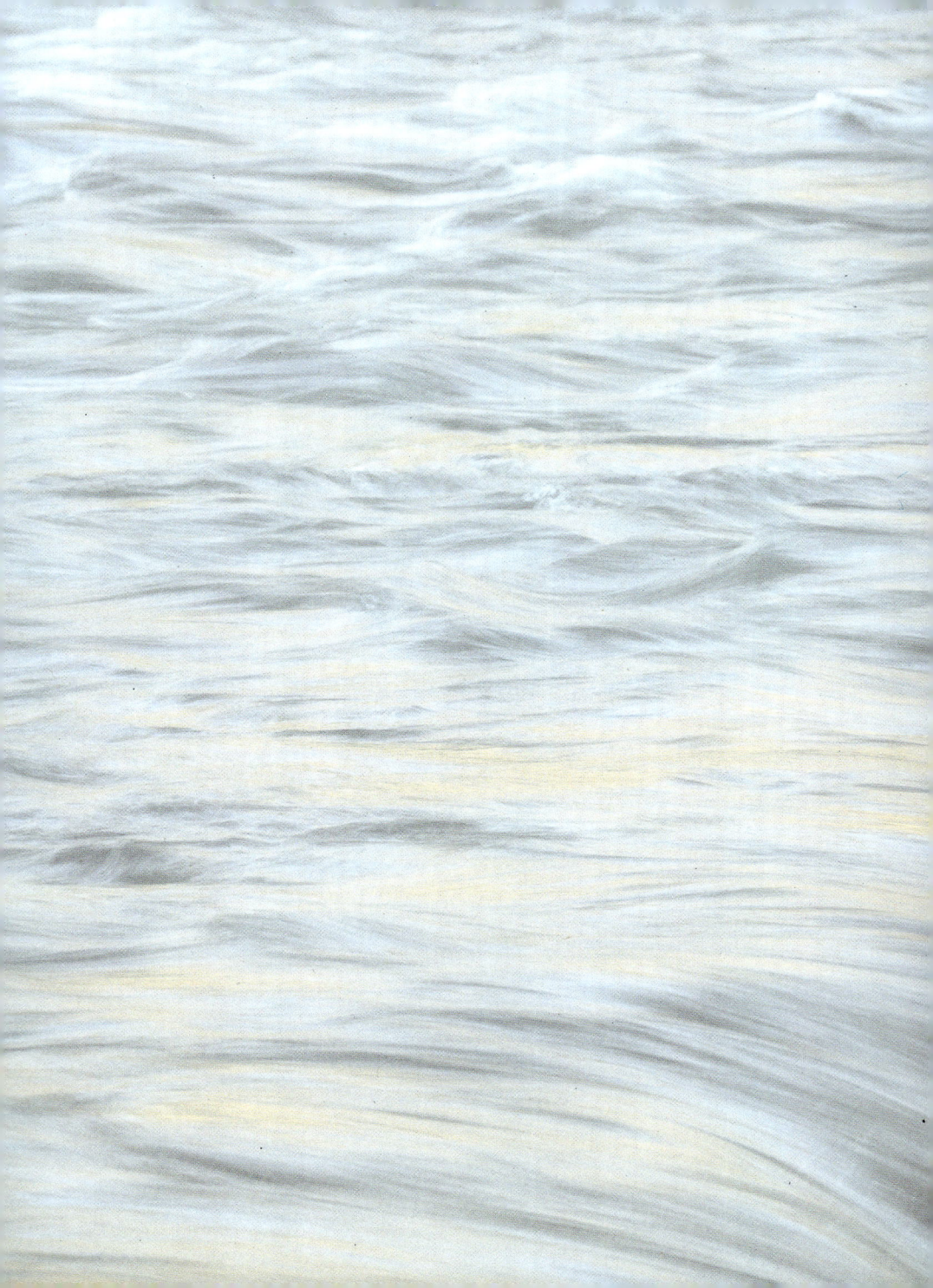

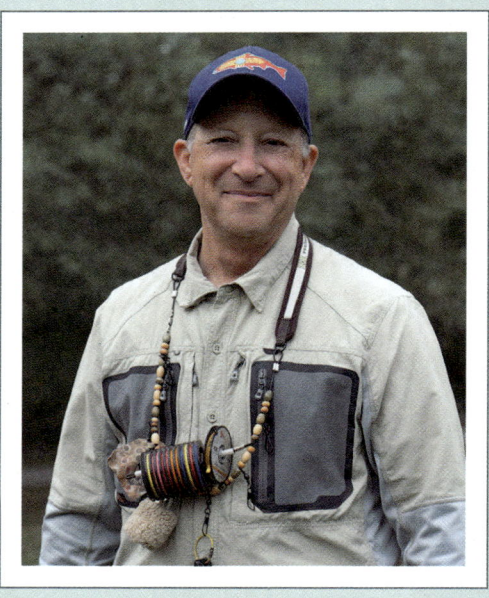

David Ladensohn is a mediator, retired executive, and entrepreneur. He was recently a Next Horizons Scholar at Oxford University. He lives in San Antonio, Texas, and outside of Santa Fe, New Mexico.